Nuclear fusion

# Nuclear fusion

**KEISHIRO NIU**

TOKYO INSTITUTE OF TECHNOLOGY

The right of the
University of Cambridge
to print and sell
all manner of books
was granted by
Henry VIII in 1534.
The University has printed
and published continuously
since 1584.

# CAMBRIDGE UNIVERSITY PRESS

## CAMBRIDGE

### NEW YORK   NEW ROCHELLE

### MELBOURNE   SYDNEY

CAMBRIDGE UNIVERSITY PRESS
Cambridge, New York, Melbourne, Madrid, Cape Town, Singapore, São Paulo, Delhi

Cambridge University Press
The Edinburgh Building, Cambridge CB2 8RU, UK

Published in the United States of America by Cambridge University Press, New York

www.cambridge.org
Information on this title: www.cambridge.org/9780521113540

Originally published in Japanese as *Kakuyugo* by
Kyoritsu Shuppan Co., Ltd, Tokyo, Japan
and © 1979 K. Niu and M. Sugiura

First published in English by Cambridge University Press 1989 as
*Nuclear Fusion*

English edition © Cambridge University Press 1989

This digitally printed version (with corrections) 2009

*A catalogue record for this publication is available from the British Library*

*Library of Congress Cataloguing in Publication data*
Niu, Keishirō, 1929–
Nuclear fusion.
Translation of: Kakuyūgō.
Includes index.
1. Nuclear fusion.    I. Title.
QC791.73.N5813 1988    539.7′64    88-16130

ISBN 978-0-521-32994-1 hardback
ISBN 978-0-521-11354-0 paperback

# Contents

Contents vii

# Contents

# Foreword to the Japanese edition

The oil crisis of 1973 generated great disturbances in the economies of advanced countries. As the economic inactivity after that crisis began to turn into prosperity, the Iranian revolution in 1978 caused a further increase in oil prices. So by the time there was an accident at an atomic reactor in the USA in the following year, the whole world had become aware of the problem of energy.

The accelerated growth in world population in the twentieth century and the recent rapid increase in energy consumption per person that has accompanied modern life styles are the fundamental causes of the present energy problem in human affairs. This problem arises not only in the fields of science and technology; it penetrates deeply into modern ways of life, economics and international relations, as well as more metaphysical concerns such as human happiness. The old idea of easily-obtainable, low-priced energy resources must be given up and a new one found to replace it. New energy resources may indeed be developed, but this will involve complex technology and long-term research by teams of scientists and engineers. And it will cost a lot of money.

This series of books was planned when the Department of Energy Sciences of the Graduate School of the Tokyo Institute of Technology was founded in Nagatsuta in 1975. It is to the authors' great pleasure that *Nuclear Fusion* is now published. The technology of nuclear fusion is incomplete at present; nor can it be deemed likely to solve the energy problem in the near future. But ultimately it may do so, and it is up to scientists and engineers to pursue this great possibility. That is why many have dedicated themselves to work on nuclear fusion.

The book consists of four chapters. In the first chapter, K. Niu sets out the fundamentals of nuclear fusion, and in the second and third chapters the principles of magnetic-confinement fusion and inertial-confinement fusion. In the final chapter, K. Sugiura discusses fusion reactors. The authors invite interested readers to indicate by direct communication where the text is incorrect or incomplete.

In closing this foreword, the authors would like to express their gratitude to the editors of the Kyoritsu Shuppan Company, who contributed to the publication of this book.

*October 1979*                                      K. Niu
                                                    K. Sugiura

# Foreword to the English edition

Energy is a source of negative entropy, in the sense that energy supplies the high-temperature thermal fluxes for human societies. As food is necessary for the growth of a human being, so is energy necessary for the development of human societies. The recent economic recession halted the increase in oil consumption, and currently the price of oil is relatively low. However, demand for oil will exceed supply, and the price of oil will rise again, when energy consumption per person in developing countries, which make up the majority of the world's population, approaches that of advanced countries. If enduring world peace at a high level of cultural development is to be realised, and by good will in accord with moral principles rather than otherwise, the world needs a source of energy whose price is both low and free from the fluctuations in supply and demand characteristic of oil.

Nuclear fusion may well provide such an energy source, and so offers us the prospect of an ultimate solution to the problem of energy. In this way, it lights a fire of hope for the future of human society. But while it holds out this great hope, nuclear fusion also presents profound scientific and technological difficulties that must be overcome before this hope can be realised. This book seeks to describe the present state and the ultimate goals of research into nuclear fusion, and to indicate some of the difficulties which lie in the path of the successful harnessing of this source of energy.

In the Japanese edition of this book, the fourth chapter, on the fusion reactor, was written by Dr Ken Sugiura, President of the Electro-Technical Laboratory in Tsukuba. Because of Dr Sugiura's heavy commitments, Dr Niu has translated that chapter, with some modifications, for publication in this English edition. Elsewhere, the principal change is in the third chapter, where there is a new section on particle-beam fusion (see K. Niu: *Frontiers in Physics Research 1*, Kyoritsu Shuppan Co, 1986, in Japanese).

The author would like to express his gratitude to Professor Heinrich Hora of the University of New South Wales, and Dr Simon Mitton of Cambridge University Press for kindly arranging the publication of this English edition of *Nuclear Fusion*. He would also like to thank Dr Erich Stuhlträger for help during his stay at the Tokyo Institute of Technology in correcting the English proofs.

*January 1987*                                           K. Niu

# 1

# Fundamentals of
# nuclear fusion

●

## 1.1. The energy problem and nuclear fusion

### 1.1.1. Energy and human societies

How can man have ruled the world, setting himself at the centre and behaving as if all other animals and plants are there for his use? There are many sayings which seek to distinguish mankind from other animals; for example, 'man is an animal which is able to laugh', or 'man is a thinking reed'. It is convenient to suppose that one effect is the direct result of just one cause. But generally speaking each effect stems from many causes. And an explanation which seems to appeal to just one concept may in fact be using that concept to stand for many others.

To say that 'man is the only animal which uses fire' is a case in point for the science and technology of energy. Although other animals are afraid of fire, man uses it; with fire he guards himself from the attack of wild animals, obtains warmth and cooks his food. We may express this more generally by saying that man is the only animal to use energy constructively in the normal course of his way of life. In their daily lives, human beings use many different kinds of energy, all of which we may represent by the single concept of 'fire'. So if we say man is what he is because he uses fire, what we are really saying is that man depends on energy.

A unit of energy, 1 Q, is equivalent to $1.05 \times 10^{21}$ J or $10^{18}$ BTU. From AD 1 to 1850, world energy consumption per years has been estimated at 0.004 Q. During this period, energy was obtained by various means: gathering fallen branches, making charcoal, collecting colza oil. The wood, charcoal and oil were supplied each year by the growth of plants and so constituted a renewable form of energy. The energy problem would not arise in the world of today if human societies still used renewable energy sources alone. In the past, man lived as he has always lived; science and technology as we know them did not exist. Nowadays we are inclined, with sentimental nostalgia, to see the life of our ancestors as happy in consequence. The reality was different. Many lived their lives in vain, at the mercy of disease; the luxury of the rich and noble few was supported by the toil of the many poor serfs.

In the eighteenth and nineteenth centuries, the industrial revolution occurred in Britain and continental Europe. Machines were substituted for human labour, and manufacturing changed from small numbers of hand-made products to mass production in factories. To begin, the energy sources substituted for human labour were wind and water power; and these, like fallen branches and charcoal, are classified as renewable energy, based on solar energy radiated to the earth.

In 1781 James Watt developed his steam engine, using coal as the heat source and transposing the motion of the piston into rotation. The use of coal as fuel brought to an end the old world based on renewable energy. Fossil fuels such as coal and oil are not renewable: they are finite stores of old solar energy locked in the remains of organic matter which grew by absorbing the radiation of the sun. Nowadays man uses only about 1 % of the energy resources of forests: 99 % of wood rots in vain, without being changed into coal or oil.

Once human societies learnt how to use non-renewable fossil fuels, world energy consumption per year averaged over the period from 1850 to 1950 increased by 0.04 Q, i.e. energy consumption was raised one order of magnitude, from 0.004 Q over the period AD 0–1850 to 0.04 Q in the next 100 years. For the 100 years from 1950–2050, average annual energy consumption is estimated to increase by another order of magnitude.

Energy is a source of negative entropy; energy sources supply human societies with heat fluxes of higher temperature. These energy fluxes are then thrown away, being released to the environment at low temperatures. In the process, however, human society increased negative entropy. Up to 1850, energy consumption was low and human society was almost closed with respect to energy. If a system is closed, it has maximum entropy and the distribution of energy is Maxwellian. Before the industrial revolution, wealth and power were distributed among the people according to such a Maxwellian distribution: the lives of the few who were rich and aristocratic were supported by their many poor retainers. After the industrial revolution, as energy consumption gradually increased, the difference in income between rich and poor decreased. In modern economically advanced countries, it now approaches equality. The energy consumption of a country may now be considered directly proportional to its gross national product.

In the twentieth century, oil has increasingly been substituted for coal. This is not because there is a shortage of coal, but rather because oil or natural gas is cleaner and more convenient. Coal deposits which can be mined economically are estimated at 100 Q, compared to 11 Q for oil. At present, annual energy consumption is estimated at 0.3 Q, and the total energy required over the next 100 years is about 70 Q. Most of this energy is currently consumed in advanced countries, but the instinctive desires of human beings for more comfortable and convenient lifestyles will increase energy consumption in developing countries. Even if advanced countries manage to save energy, the energy needed over the next 100 years cannot be supplied by oil, a fact already reflected in the dependence of the price of oil on international relations.

In the twentieth century mankind has been able to rely on energy supplied by the combustion of oil, but if we continue to do so in the face of the finite stock of this unrenewable supply of energy, social confusion will result. So if mankind is to maintain a civilised lifestyle in the future, new energy sources must be developed within the next 100 years.

The energy given out by the sun is enormous; only a fraction of it (but still a great amount) reaches the earth. Our ancestors used the sun as the source of energy for agriculture. Can we sustain civilised life for the forseeable future by using the sun's energy, rather than fossil fuels, as the energy source for giant industries? The sun radiates energy at $9.3 \times 10^{21}$ kcal per second, $4.1 \times 10^{13}$ kcal per second reaches the earth, $5 \times 10^{-9}$ of the total. Over a year, the earth receives $1.3 \times 10^{21}$ kcal, or $5 \times 10^3$ Q per year. If we consume 1 Q per year, the energy received from the sun per year corresponds to 5000 years annual consumption; the amount consumed is thus very small compared to the amount received. However, the density of the sun's energy on a plane normal to the earth's surface is as low as 2 cal/cm$^2$ min, assuming none is absorbed between the sun and the earth. Because of this low energy, a large area is needed in order to collect a significant amount of the sun's energy. Generally speaking, it is difficult to utilise low-density energy. When a power station is constructed, the energy pay rate, here defined as

energy pay rate

$$= \frac{\text{total amount of energy generated during operational life}}{\text{total energy consumed in constructing the station}}$$

must be greater than one, from the technological point of view (leaving aside the economics of the matter). Operational life is usually taken to be 20 years. The total energy consumed includes mining the iron ore, turning it into steel, transporting to the site and then constructing and commissioning the plant. Fuel energy – consumed in extraction, transportation and treatment of waste products – must also be included. If a large power supply system is to be constituted by the same kind of station, not one but many stations must be constructed in succession: one station is needed to create the energy by which another is constructed during its operational life. Consequently, the total energy in the denominator of the pay rate equation is frequently doubled. The pay rate will seldom be greater than one if the station uses low-density energy as input; the station must be large, so the value of the denominator becomes large too.

When low-density energy is used, it is desirable to operate the station at low temperatures: water at 40°C can be used for heating, and water at 75°C for cooling. But it is difficult to make the energy pay rate greater than one when solar energy is used to heat the water to such temperatures. Needless to say, the energy pay rate will never exceed one if solar energy is to provide high temperatures, such as those for the generation of electricity (1400°C for an oil-powered station, 800°C for a nuclear one). Natural (soft) energy sources are out of the question as far as high-output stations are concerned.

### 1.1.2. Energy in classical mechanics

In classical mechanics, the kinetic energy of a body of mass $m$ with velocity $v$ is given by

$$KE = \tfrac{1}{2}mv^2. \tag{1.1}$$

If the body is located in the field of a conservative force, such as gravity, it will have a potential energy given by

$$U = mgh \tag{1.2}$$

where $h$ is the height above ground level in a field of gravitational strength $g$. The sum of the kinetic energy and the potential energy is the total energy $E$, and by the law of conservation of energy,

$$E = KE + U = \tfrac{1}{2}mv^2 + mgh = \text{constant}. \tag{1.3}$$

If the body is released at height $h$, its potential energy decreases and kinetic energy increases as it falls; at ground level, all its energy is in the form of kinetic energy. In order to extend the law of conservation of energy to what happens once the body hits the ground, the concept of heat energy must be introduced. The heat energy $J$ absorbed by the ground after impact is equal to the total energy $E$. The mass of the body is constant, whether it is at rest at height $h$, falling, or hitting the ground; in classical mechanics, energy and mass are conserved separately.

### 1.1.3. Chemical energy

All matter consists of molecules, the ultimate units of chemical character. A molecule consists of atoms, and atoms consist of a nucleus with a positive electric charge and electrons with a negative charge. The greater part of the mass of an atom is in the nucleus: the mass of an electron is much smaller than that of a nucleus. Further, the nature of the atom depends on the nucleus: if its mass or charge is different, it is a different atom. The chemical nature of an atom, however, depends solely on the charge of the nucleus. When the charges of two nuclei are the same but the masses are different, they are called isotopes and are classified as the same atom. For the complete atom, the nucleus is said to have a charge $Ze$, where $e$ is the charge of an electron and $Z$ is a positive integer known as the atomic number, and $Z$ electrons, each with charge $-e$, revolving around the nucleus. The chemical energy of an atom stems from the potential energy of these electrons.

Suppose a molecule $A$ consists of an atom $X$ and an atom $Y$. The nuclei of atoms $X$ and $Y$ are neutralised by electrons when they form molecule $A$. If the two nuclei approach each other within a distance of less than $10^{-10}$ m, the electrons (which revolve separately around their own nucleus when the nuclei are distant from one another) change their orbits to ones that revolve around the two nuclei together. If the potential energy of the electrons moving in the new orbits is smaller than the sum of the potential energies of the electrons separated into two atoms, the two atoms will constitute one molecule $A$. Take the chemical reaction

$$A + B \rightarrow C + E_c \tag{1.4}$$

where $A$, $B$ and $C$ are molecules. The reaction proceeds because the sum of the electron energies of the two molecules $A$ and $B$ is larger than that of the molecule $C$. The difference $E_c$ in the energies of the electrons is released and called the heat of reaction.

In the twentieth century, human beings have used energy from the combustion of coal and oil. This thermal energy has come from the changes in the potential energy of electrons revolving around nuclei resulting from reactions such as eq. (1.4).

### *1.1.4. Atomic energy*

In 1945, atomic energy was released by the atomic bombs dropped on Hiroshima and Nagasaki. From the mid-1950s atomic energy could be used for peaceful purposes. Atomic energy and chemical energy have quite different origins; the theoretical foundations of atomic energy were laid by 'the special theory of relativity' developed by Albert Einstein in 1905.

In classical mechanics, mass, momentum and energy are conserved separately before and after interactions. The total energy of a body of mass $m$, according to the special theory of relativity, is given by

$$E = \frac{m_0 c^2}{\sqrt{1 - v^2/c^2}}. \tag{1.5}$$

In eq. (1.5), $m_0$ is the rest mass of the body and $c$ is the speed of light in a vacuum. The mass of the body is not constant but is related to its velocity $v$ by

$$m = \frac{m_0}{\sqrt{1 - v^2/c^2}}. \tag{1.6}$$

Equation (1.5) reduces to

$$E = m_0 c^2 + \tfrac{1}{2} m_0 v^2, \tag{1.5'}$$

if $v/c$ in eq. (1.5) is expanded as a series, provided that $v$ is smaller than $c$. The second term on the right-hand side of eq. (1.5') is the kinetic energy, while the first term is a constant, called the rest energy. Because there is a term in eq. (1.5') which includes the rest mass $m_0$ for the energy $E$, mass and energy are not conserved separately, but are related to each other. Equation (1.5') shows that a new energy $m_0 c^2$ must appear when the rest mass $m_0$ disappears. A mass of 1 kg corresponds to an energy of $9 \times 10^{16}$ J, which is quite a large amount. The mass of the nucleus decreases when the mass changes into energy, because the majority of the mass of the atom is located in the nucleus. In chemical reactions, the nucleus does not change. Atomic energy, on the other hand, uses the change in mass of the nucleus.

### *1.1.5. The Coulomb force and the nuclear force*

Let us investigate nuclei in more detail. A nucleus is made up of more fundamental constituents called elementary particles. Although many elementary particles have been observed up to now, the electron, proton and neutron are the most important ones.

An electron has a mass of $9.10988 \times 10^{-31}$ kg and charge of $-1.6 \times 10^{-19}$ C, and revolves around the nucleus. The nucleus consists of protons, whose mass is $1.673 \times 10^{-27}$ kg and charge is $1.6 \times 10^{-19}$ C, and neutrons, whose mass is $1.675 \times 10^{-27}$ kg and charge is zero. Protons, together with neutrons, are called nucleons. A hydrogen atom has one electron, which revolves around a nucleus consisting of a single proton. In the case of a helium atom, two electrons revolve around a nucleus consisting of two protons and two neutrons. A helium nucleus is expressed as $^4_2\text{He}$. The upper-left number is the mass number (the mass of the helium nucleus has four times the mass of the hydrogen nucleus, and the mass number is the number of nucleons that constitute the nucleus) and the lower-left number is the atomic number or the charge number (the charge of the helium nucleus has twice the charge of a proton nucleus and the charge number is the number of protons in the nucleus).

The neutron is electrically neutral while the proton has a positive charge. A repulsive force, called the Coulomb force, acts on any two particles with the same kind of charge. The force $F$ acting on two bodies whose charges are respectively $Z_1e$ and $Z_2e$ is

$$F = \frac{Z_1Z_2e^2}{4\pi\varepsilon r^2}. \tag{1.7}$$

Here $\varepsilon = 8.854 \times 10^{-12}$ F/m is the dielectric constant in a vacuum. The potential energy $U$ of the two charges whose distance is $r$ is given by

$$U = -\int_\infty^r \frac{Z_1Z_2e^2}{4\pi\varepsilon r^2}\, \mathrm{d}r = \frac{Z_1Z_2e^2}{4\pi\varepsilon r}. \tag{1.8}$$

Here the potential energy is chosen to be zero at infinity. Take any two protons. At $r_0 = 5 \times 10^{-15}$ m, which is the radius of a nucleus, $U = 0.29$ MeV. Here $1\,\text{eV} = 1.602 \times 10^{-19}$ J is the kinetic energy obtained by an electron which is accelerated by an electric potential of 1 V. When the distance between the centres of the two nuclei becomes less than $5 \times 10^{-15}$ m, a strong attractive force called the nuclear force acts on the nuclei. This nuclear force binds protons and neutrons together in a nucleus. Figure 1.1 shows the relation between the potential energy $U$ and the distance $r$ between the two protons. Potential energy increases by the Coulomb force as $r$ decreases from $\infty$ to $r_0$. Due to the nuclear force, potential energy decreases in the region within $r_0$. $U_0$ in Fig. 1.1 is $U = 0.29$ MeV for two protons at $r = r_0$, as given by eq. (1.8). The value of $U_1$ is negative because the potential energy due to the nuclear force is larger than that due to the Coulomb force. Two protons become stable when they form one nucleus, rather than two separate protons at $r = \infty$.

Instead of two protons, let us now consider two nuclei, whose charges are $Z_1 e$ and $Z_2 e$ respectively. The Coulomb force between the nuclei is expressed as $F = Z_1 Z_2 e^2 / 4\pi \varepsilon r^2$. On the other hand, the distance $r_0$ inside which the nuclear force acts, and the nuclear force itself for these two nuclei, are not much different from those for the two protons. In other words, the nuclear force does not depend much on $Z$. Thus the potential energy for these two nuclei varies as shown in Fig. 1.2. The value of $U_0'$ in Fig. 1.2 is $Z_1 Z_2$ times that of $U_0$ in Fig. 1.1. As the difference between $U_0 - U_1$ and $U_0' - U_1'$ is not large, $U_1'$ becomes positive (see Fig. 1.2) if $Z_1 Z_2$ is large. In

**Fig. 1.1.** Potential energy due to the Coulomb force and the nuclear force between two protons.

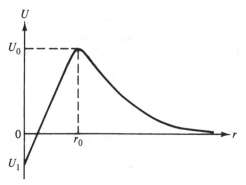

**Fig. 1.2.** Potential energy due to the Coulomb force and the nuclear force between two nuclei with large $Z$.

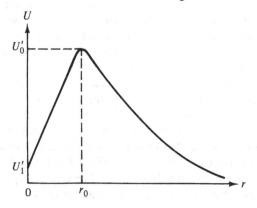

such a case, the two separate nuclei at $r = \infty$ are more stable than one large nucleus formed by the two nuclei. In summary, one large nucleus becomes stable when it is separated into two, while two small nuclei become stable when they are combined to form one large nucleus. The nucleus of $Z = 26$, i.e. $_{26}$Fe, at the centre of the periodic table, is the most stable.

### 1.1.6. The nuclear fission reaction

The uranium nucleus is large and unstable. Because it has no charge, a neutron can approach a nucleus easily. If a uranium nucleus absorbs a neutron, the nucleus becomes unstable, resulting in nuclear fission. The typical fission reaction of uranium is given by

$$^{235}_{92}U + ^{1}_{0}n = 2 \times ^{118}_{50}Sn + 8e + 236 \text{ MeV.} \qquad (1.9)$$

Here n is the neutron and e is the electron. Equation (1.9) describes a uranium nucleus which is split into two equal Sn nuclei. The atomic weight of $^{235}_{92}U$ is 235.124 (1 atomic weight unit is $1.6603 \times 10^{-27}$ kg), i.e. the atomic mass of $^{235}_{92}U$ is $3.903764 \times 10^{-25}$ kg. As the atomic weight of $^{118}_{50}Sn$ is 117.94, the atomic mass of $^{118}_{50}Sn$ is $1.958158 \times 10^{-25}$ kg. The mass of $-4.201 \times 10^{-28}$ kg is the difference between double the mass of tin and the sum of the masses of uranium and of a neutron. This is the mass deficiency in the nuclear fission of $^{235}_{92}U$ and corresponds to an energy of $3.78 \times 10^{-11}$ J $= 236$ MeV. An atomic power station uses this mass deficiency to generate electricity.

In 1951, JRR-1, Japan's first fission reactor, began operation. When the reactor reached the critical state, the radio announcer said 'a second fire now burns in our country'. The 'first fire' is the combustion through which man used energy in his daily life. Atomic energy is the second fire, newly obtained. The atomic energy extracted from light-water reactors uses the mass deficiency in the fission of uranium nuclei.

The amount of uranium in mineral deposits is estimated at $2.5 \times 10^{10}$ kg. Two isotopes $^{235}_{92}U$ and $^{238}_{92}U$ are found in natural uranium. Only $^{235}_{92}U$ undergoes nuclear fission in the light-water reactor; natural uranium includes 0.7 % of $^{235}_{92}U$. The thermal efficiency of the atomic power station is about 20 %. This low efficiency comes from the safety margin of the reactor, which operates at a low temperature. In light-water reactors, electric energy of only 7 Q can be released by use of the natural uranium of $2.5 \times 10^{10}$ kg. Atomic energy as the 'second fire' cannot support human civilisation for long, if we take into account that human beings will use 1 Q of energy per year in the near future.

In seawater, uranium is found in concentrations of 0.0033 ppm. The total amount is $4.4 \times 10^{12}$ kg, which is 200 times that of the uranium as a mine resource. If we develop techniques to collect the dilute uranium from seawater, nuclear fission energy could help us to sustain our civilised lifestyle for a longer time.

When $^{238}_{92}U$ absorbs a high-speed neutron, $^{239}_{92}U$ is changed to $^{239}_{94}Pu$ as

$$^{238}_{92}U + {}^{1}_{0}n \rightarrow {}^{239}_{92}U \rightarrow {}^{239}_{93}Np \rightarrow {}^{239}_{94}Pu. \qquad (1.10)$$

Plutonium undergoes nuclear fission too. In the fast breeder reactor, $^{238}_{92}U$ is reacted with $^{235}_{92}U$ and changed to $^{239}_{94}Pu$. Taking into account that the breeding rate (change rate of $^{238}_{92}U$ to $^{239}_{94}Pu$) is 60 % and the thermal efficiency can be increased to 30 % with increase in safety (due to the improvement of the wall materials), we can expect that a fission energy of 300 Q can be extracted from the uranium as a mine resource. It is hoped that the technology of the fast breeder reactor will soon be perfected (the Super-phoenix of 1.2 GW is already operating in France), together with a fuel cycle which includes the extraction of unburnt uranium and plutonium.

The nuclear fission reactor can confine fission products with strong radio-activities (plutonium is especially dangerous to human beings; it is the explosive of the atomic bomb) in the fuel rod; by contrast, a coal power plant exhausts polluted materials into the atmosphere. The safe management and utilisation of a large amount of radio-active fission products over a longer period is the most important and most difficult technological task involved in using the fission reactor.

### 1.1.7. Energy from nuclear fusion

The sun is a giant sphere with radius $7 \times 10^8$ m and mass $2 \times 10^{30}$ kg. It radiates $9.3 \times 10^{21}$ kcal of energy per second; it would have burnt itself out in 3000 years if it had been burning coal. Before Einstein many wondered why the sun continues to shine. The special theory of relativity explained its enduring life.

The nucleus of hydrogen is a single proton. When it combines with other protons to form a larger nucleus, it becomes more stable. Consider the following nuclear reaction:

$$4^1_1H + 2e = {}^4_2He + 27.05 \text{ MeV}. \qquad (1.11)$$

The atomic weight of $^1_1H$ is 1.008 while the atomic weight of $^4_2He$ is

4.003. The mass decreases by 7 % in reaction (1.11), i.e. the mass deficiency is $4.815 \times 10^{-29}$ kg, which corresponds to an energy of $4.3334 \times 10^{-12}$ J = 27.05 MeV. The sun continuously shines because of this nuclear fusion. By the fusion reaction given by (1.11), the sun loses 1/3000 of its mass over $5 \times 10^9$ years. It will be to our good fortune if we can utilise reaction (1.11) on the earth. We have enough hydrogen on earth, and the fusion product, helium, is a stable inert gas which is useful and harmless. However, from a small power station on the earth (by contrast to the giant sun), the fusion energy released in reaction (1.11) cannot be obtained because of the very low reaction rate.

The hydrogen found on earth includes deuterium ($^2_1$H, i.e. $^2_1$D) in the mass ratio of 1:5000. Deuterium undergoes the following two types of fusion reaction, with the same reaction rate:

$$^2_1\text{D} + {}^2_1\text{D} = \begin{cases} ^3_2\text{He} + {}^1_0\text{n} + 3.27 \text{ MeV}, & (1.12) \\ ^3_1\text{T} + {}^1_1\text{H} + 4.03 \text{ MeV}, & (1.13) \end{cases}$$

where $^3_1$T is tritium and is an isotope of hydrogen. The reaction rate of (1.12) or (1.13) is much higher than that of (1.11). A human being uses in his lifetime an energy of $2 \times 10^{16}$ J, which can be extracted from a bathtub full of heavy water ($D_2O$ of 1 t) by using the fusion energy. An oilcan full of heavy water will supply enough energy to run a car for a lifetime if the car has a nuclear-fusion engine. On the earth, there is $10^{20}$ l of water, in which deuterium can supply fusion energy of $7.5 \times 10^9$ Q. This is enough energy for $10^{10}$ years for human needs. Given present scientific knowledge and technology, however, the reaction cross-section of (1.12) or (1.13) is not large enough to enable that fusion energy to be extracted. A fusion reaction with a larger cross-sectional area,

$$^2_1\text{D} + {}^3_1\text{T} = {}^4_2\text{He} + {}^1_0\text{n} + 17.6 \text{ MeV}, \qquad (1.14)$$

has the greatest possibility of being utilised. In eq. (1.14), T is a radio-active material which undergoes $\beta$-decay with the half-life of 12 years, so there is no natural tritium. But tritium can be produced via the reactions

$$^6_3\text{Li} + {}^1_0\text{n} = {}^4_2\text{He} + {}^3_1\text{T} + 4.8 \text{ MeV}, \qquad (1.15)$$

$$^7_3\text{Li} + {}^1_0\text{n} = {}^4_2\text{He} + {}^3_1\text{T} + {}^1_0\text{n} - 2.47 \text{ MeV}. \qquad (1.16)$$

In fusion reactors, reaction (1.14) proceeds at the same time as (1.15) or (1.16). Neutrons produced by reaction (1.14) are supplied to either or both reactions (1.15) and (1.16); and tritium produced by either or both reactions (1.15) and (1.16) is used in reaction (1.14). If reactions (1.14) and (1.15) are combined to give

$$\text{{}_3^6\text{Li}} + {}_1^2\text{D} = 2{}_2^4\text{He} + 22.4 \text{ MeV}, \qquad (1.17)$$

it becomes clear that the real fuels for the D–T reaction are lithium and deuterium.

Although tritium undergoes $\beta$-decay with the half-life of 12 years, its radio-activity is weak. Since tritium is an isotope of hydrogen, tritium reacts chemically with oxygen and becomes water. If we drink the water which includes tritium, the water remains in our body for some time; the tritium can affect the reproductive organs via its $\beta$-rays. Because tritium leaks from vessels easily, it must be carefully controlled when the D–T reaction is used in the reactor.

The world's lithium mines are located principally in the USA, Canada and a number of African countries. The total amount of Li as a mine resource is of the order of $9 \times 10^9$ kg, which can yield fusion energy of 3000 Q via reaction (1.17). Seawater includes D at 158 ppm, as well as Li at 0.17 ppm. The total amount of Li in seawater corresponds to fusion energy of $6 \times 10^6$ Q. Although Na, K, Ca, Mg, Be and similar elements, as well as Li, are dissolved in seawater at higher concentrations, techniques of separating Li will soon be developed for Li is the smallest alkali metal.

If we limit ourselves to the D–T reaction only, the fusion energy of $6 \times 10^6$ Q is sufficient for present human needs. Nuclear fusion is a golden casket, large enough to solve the energy problem completely. Accordingly, God has not given it to us easily; mankind is now trying, with appropriate pains, to find it.

The energy problem is not limited to science and technology. It seems that the human sciences follow the natural sciences, even now. The energy problem will greatly affect the human sciences too, through the revolution in modes of thinking about the food problem, North–South relations, international relations, and economic and political problems. Enough energy supply will lead us to an ideal society with a decreasing entropy.

## 1.2. The nuclear fusion reaction

### 1.2.1. Thermal motion

To collide two protons with each other so as to cause a nuclear fusion reaction, the kinetic energy $KE$ due to the relative velocity of the protons must be larger than the potential energy $U_0$ of the Coulomb force, because the two protons must approach each other in opposition to the Coulomb force within $r_0$, the distance inside which the nuclear force acts. How can nuclei obtain the kinetic energy needed to cause nuclear fusion reactions? The amount of energy released by one fusion reaction is small. It is necessary that many nuclei successively undergo nuclear fusion reactions during some period if a usable amount of energy is to be extracted from fusion reactions as atomic power. One method is to imitate the sun and to heat nuclei (strictly speaking, the gas must be a plasma, which has electrons to neutralise the charge of nuclei; otherwise the nuclei will be scattered by the Coulomb force). When the temperature of a gas increases, its molecules dissociate into atoms at some particular temperature. As the gas continues to be heated, the electrons at last free themselves from the restraints of the nuclei around which they rotate. A gaseous material that consists of free nuclei with positive charges and free electrons with a negative charge is called a plasma. The temperature at which a gas is ionised depends on its atoms. Roughly speaking, almost all gases become plasma above $10^6$ K. For a gas at the temperature $T$, we have a relation as follows,

$$\tfrac{3}{2}kT = \tfrac{1}{2}m\bar{v}^2, \tag{1.18}$$

where $m$ is the mass of the nucleus, $\bar{v}$ is the thermal velocity and $k = 1.3709 \times 10^{-23}$ J/deg is the Boltzmann constant; thus we have

$$1 \text{ eV} = 1.602 \times 10^{-19} \text{ J} = 1.16 \times 10^4 \text{ K}.$$

If the kinetic energy of a nucleus with velocity $v$, given by eq. (1.18), is to exceed the potential energy $U_0 = 0.29$ MeV of the Coulomb force, the temperature of the plasma must be higher than $T_0 = 0.29 \times 10^6 \times 1.16 \times 10^4$ K $= 3.364 \times 10^9$ K.

### 1.2.2. The equilibrium distribution function

It is not true to say there are no fusion reactions in hydrogen plasma at a temperature less than $T_0$. When the plasma is in an equilibrium state, the velocity distribution function $f$ for the proton is given by the Maxwellian equation

$$f = n\left(\frac{m}{2\pi\kappa T}\right)^{3/2} \exp\left\{-\frac{m}{2kT}v^2\right\}. \tag{1.19}$$

Here $m$ is the proton mass and $n$ is the number density. We should understand that the velocities of protons are distributed over a wide range, although those with $\bar{v}$ given by eq. (1.18) are in the majority. There are protons which have velocities greater than $\bar{v}$, though their number decreases exponentially with velocity. Accordingly, there may be protons whose kinetic energies exceed $U_0$ at a temperature lower than $T_0$. It is possible for these protons to cause fusion reactions.

### 1.2.3. The tunnel effect

In section 1.2.1 it was said that the kinetic energies of nuclei with respect to their relative motions should always exceed the Coulomb potential. According to quantum mechanics, however, a nucleus has a chance of approaching other nuclei across $r_0$, penetrating the barrier of the Coulomb potential. This phenomenon is called the tunnel effect.

In conjunction with the fact described in the preceding section, protons whose temperature is lower than $T_0$ can cause fusion reactions. The number of reactions, however, depends significantly on temperature.

### 1.2.4. Cross-sectional area and mean free path

Let us investigate the number of fusion reactions in a plasma whose number density is $n$ and temperature is $T$. The nucleus is assumed to be a hard sphere of diameter $d$, and the number of chain-collisions of one nucleus with other nuclei per unit time is counted. Although the nucleus changes its path line at each collision, the distance the nucleus moves in unit time is $\bar{v}$, given by eq. (1.18), if the refracted path line is stretched out to a straight line. The number $v$ of collisions during the period in which the nucleus moves a distance $\bar{v}$ is equal to the number of nuclei whose centres are located in a cylinder whose diameter is $2d$ and height is $\bar{v}$, as shown in Fig. 1.3. Since the number density is $n$,

$$v = \pi\bar{v}d^2n. \tag{1.20}$$

The value $\sigma = \pi d^2$ is the area of the target if the collided nucleus is assumed to be the target (the area of the colliding nucleus is also taken into consideration), and is called the collision cross-sectional area. The collisional cross-sectional area has the dimension of $m^2$.

**Fig. 1.3.** The volume removed by the motion of one nucleus of diameter $d$ in unit time.

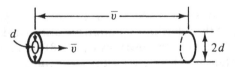

The unit 1 barn $= 10^{-28}$ m$^2$ is frequently used for $\sigma$, since $\sigma$ has a small value. With $\sigma$, eq. (1.20) reduces to

$$v = \sigma \bar{v} n. \tag{1.21}$$

The mean free path is defined as the mean distance the nucleus moves without collision. So we have

$$l = \bar{v}/v = 1/\sigma n \tag{1.22}$$

where $l$ is the mean free path. The number $v$ in eq. (1.21) is the number of collisions which one nucleus experiences in unit time. Because there are $n$ nuclei per unit volume, the collision frequency $N$ which is the number of collisions in unit time and in unit volume seems to be $\sigma \bar{v} n^2$. But this number comes from a counting method by which one collision is counted twice by collided and colliding nuclei. Thus the real number is

$$N = \tfrac{1}{2}\sigma \bar{v} n^2. \tag{1.23}$$

In eq. (1.23), the number of collisions are counted for the nuclei of a rigid sphere. In the case of real nuclei, some nuclei undergo Coulomb scatterings when they approach each other, and some nuclei undergo nuclear fusion. The ratio of nuclei which undergo fusion reactions depends on the relative velocity in collisions resulting from the tunnel effect. If we define the cross-sectional area $\sigma$ of fusion reaction taking its probability into account, the number $N$ for real nuclei can also be expressed by eq. (1.23). In a plasma with a temperature $T$, $\bar{v}$ in eq. (1.23) is replaced by $v$, and $N$ is averaged with respect to $v$ by using the Maxwellian distribution, since $\sigma$ depends on the velocity. $\langle \sigma v \rangle$ denotes the product of $\sigma v$, where $v$ is averaged by Maxwellian distribution. We then have

$$N = \tfrac{1}{2}\langle \sigma v \rangle n^2. \tag{1.24}$$

### 1.2.5. Important fusion reactions

A usable amount of fusion energy is yielded if $N$ is a large number. A large $N$ can be obtained by a large $n$ as well as by a large $\langle \sigma v \rangle$. The value of $\langle \sigma v \rangle$ depends on the temperature and the kind of fusion reactions. It is advantageous to use fusion reactions with a large $\langle \sigma v \rangle$, because a dense and high-temperature plasma is usually difficult to obtain on earth. The nuclear reaction of the first generation is

$$^2_1\text{D} + {}^3_1\text{T} = {}^4_2\text{He} + {}^1_0\text{n} + 17.6 \text{ MeV}, \tag{1.25}$$

and that of the second generation is

$$\mathstrut^2_1 D + \mathstrut^2_1 D = \begin{cases} \mathstrut^3_2 He + \mathstrut^1_0 n + 3.27 \text{ MeV}, & (1.26) \\ \mathstrut^3_1 T + \mathstrut^1_1 H + 4.03 \text{ MeV}. & (1.27) \end{cases}$$

$\mathstrut^2_1 D$ consists of one proton and one neutron. $\mathstrut^3_1 T$ consists of one proton and two neutrons. The collision frequency $N$ for the D–D reactions (1.26) and (1.27) is given by (1.24). The values of $\langle \sigma v \rangle$ for both the reactions are almost equal with each other. On the other hand, D must collide with T in the D–T reaction (1.25). If we denote the number densities of D and T by $n_D$ and $n_T$ respectively, then $N$ for the D–T reaction (1.25) becomes

$$N = \langle \sigma v \rangle n_D n_T. \qquad (1.28)$$

When D and T have the same number density $n_D = n_T = n/2$, eq. (1.28) reduces to

$$N = \tfrac{1}{4} \langle \sigma v \rangle n^2. \qquad (1.29)$$

The values $\langle \sigma v \rangle$ for the D–D reactions (the sum of $\langle \sigma v \rangle$ of eqs (1.26) and (1.27)) and for the D–T reaction are shown in Fig. 1.4 as a function of the temperature $T$. The collision rate $\langle \sigma v \rangle$ is very small below $T = 1$ keV. The value of $\langle \sigma v \rangle$ for the D–D reactions is smaller by two orders of magnitude by comparison with $\langle \sigma v \rangle$ for the D–T reaction at $T = 10$ keV. Therefore our immediate aim is to realise the D–T reaction.

Deuterium exists in the natural hydrogen in the weight ratio of $1 : 5000$. Although T is necessary for the D–T reaction, no T exists

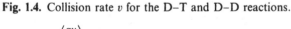

**Fig. 1.4.** Collision rate $v$ for the D–T and D–D reactions.

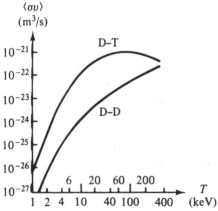

in nature, because T undergoes $\beta$-decay with a half-life of 12 years. T is produced by using lithium, as follows:

$$^6_3\text{Li} + ^1_0\text{n} = ^4_2\text{He} + ^3_1\text{T} + 4.8 \text{ MeV}, \tag{1.30}$$

$$^7_3\text{Li} + ^1_0\text{n} = ^4_2\text{He} + ^3_1\text{T} + ^1_0\text{n} - 2.47 \text{ MeV}. \tag{1.31}$$

The exothermic reaction (1.30) can contribute to the energy output of the reactor. On the other hand, one D–T reaction produces one neutron, as eq. (1.25) shows. According to reaction (1.30), one tritium nucleus is produced per neutron. If there are neutrons which do not react with lithium, the quantity of the tritium decreases. Although eq. (1.31) shows an endothermic reaction, another neutron is produced by the reaction. The ratio of $^7\text{Li}$ in the fuel must be regulated in order that the quantity of the tritium does not decrease (in other words, in order that the breeding ratio of T exceeds unity). In the case that the number of neutrons in the reactor becomes too low, beryllium is added to lithium to utilise a neutron-multiplying reaction

$$^9_4\text{Be} + ^1_0\text{n} = 2^4_2\text{He} + 2^1_0\text{n}. \tag{1.32}$$

### 1.2.6. Energy in fusion reactions

In the preceding section it was said that a kinetic energy equal to or greater than a potential energy of 0.29 MeV (the potential energies for D and T are the same as for H because D and T are the isotopes of H) is required to enable nuclei to undergo fusion reactions against the Coulomb force. But this kinetic energy is negligibly small in comparison with the reaction energy of 17.6 MeV for D–T reaction. The reaction energy of 17.6 MeV is the sum of the kinetic energies of an $\alpha$-particle and a neutron which are the fusion products. Momentum must be conserved before and after the reaction. If we assume the small sum of the momenta before the reaction to be zero, the ratio of the velocities of the $\alpha$-particle to that of the neutron becomes $1:4$, inversely proportional to the mass ratio. That is, the kinetic energy of the $\alpha$-particle is 3.5 MeV, which is $\frac{1}{5}$ of 17.6 MeV, and that of the neutron is 14.1 MeV, which is $\frac{4}{5}$ of 17.6 MeV. A similar consideration leads us from reaction (1.26) to the result that $^3_2\text{He}$ has 0.82 MeV and $^1_0\text{n}$ has 2.45 MeV. For reaction (1.27), $^3_1\text{T}$ has 1.01 MeV and $^1_1\text{H}$ has 3.02 MeV. Let us assume that the temperature of the D–T plasma is 10 keV. The collision rate $\langle \sigma v \rangle$ is $1.1 \times 10^{-22}$ m$^3$/s at this temperature. By using eq. (1.29), the fusion energy extracted from the plasma per second and per m$^3$ is given by

$$P_\text{f} = \tfrac{1}{4}\langle \sigma v \rangle n^2 E_\text{f} = 8 \times 10^{-35} n^2 \text{ W/m}^3, \tag{1.33}$$

where $E_f = 17.6$ MeV is the reaction energy. If we choose $P_f = 10^8$ W/m³, which is a comparable output power density in a fission reactor, the number density $n$ of the plasma must be

$$n = 1.16 \times 10^{21}/\text{m}^3. \tag{1.34}$$

### 1.2.7. Advanced nuclear fusion reactions

Strong efforts are being made to obtain energy utilising the D–T reaction, which has the largest $\langle \sigma v \rangle$ at low temperatures. In this D–T reaction, however, lithium is used because tritium does not exist in nature. If possible, it would be preferable to use the D–D reactions, because deuterium can be obtained easily from seawater. There is a possibility that the atmosphere will be contaminated by the tritium that is produced in the D–D reactions. In addition, reactions (1.25)–(1.27) produce neutrons, which have much of the reaction energy. These neutrons do not contribute to heating the plasma; since neutrons have zero charge, they escape from the plasma immediately. There is no direct way to convert their kinetic energy to electrical energy because of their electrical neutrality. These neutrons are also harmful to human beings if they leak from the reactor, which they may do because the collision cross-section with the wall material is small. The plasma must be surrounded by a protective wall of a thick and heavy material which is a good neutron-absorber.

If the following third-generation nuclear reactions can be utilised,

$$^1_1\text{H} + ^{11}_5\text{B} = 3^4_2\text{He} + 8.7 \text{ MeV}, \tag{1.35}$$

$$^1_1\text{H} + ^6_3\text{Li} = ^3_2\text{He} + ^4_2\text{He} + 4.0 \text{ MeV}, \tag{1.36}$$

the only fusion products are charged particles which are not radio-active. A high rate of energy conversion can be expected from these reactions if a direct conversion method is used. For these reactions, $\langle \sigma v \rangle$ at low temperatures is very small. An electron temperature of 140 keV, and an ion temperature of 280 keV, are required to obtain fusion energy from the reaction (1.35).

## 1.3. The Lawson criterion

### 1.3.1. Heating of plasma by fusion reactions

Equation (1.33) gives the fusion energy released per unit volume and unit time. This fusion energy is shared by the kinetic energies of neutrons and of α-particles, the fusion products. Neutrons pass through and escape from the plasma without interacting with charged particles in it because of their electrical neutrality. In the energy given by eq. (1.33), the amount of the energy $P_\alpha$ that is absorbed by the plasma, thus heating it, is

$$P_\alpha = \tfrac{1}{4}\langle \sigma v \rangle n^2 E_\alpha, \tag{1.37}$$

in which the kinetic energy $E_\alpha = 3.5$ MeV of an $\alpha$-particle replaces $E_f$ in eq. (1.33). If $\langle \sigma v \rangle = 1.1 \times 10^{-22}$ m$^3$/s at $T = 10$ keV is inserted in eq. (1.37), then $P_\alpha = 1.6 \times 10^{-35} n^2$ W/m$^2$, which is the heat added to the plasma per unit volume and per unit time by the fusion reactions.

### 1.3.2. Bremsstrahlung

Plasma has a tendency to decrease in temperature due to the outflowing of energy. The largest part of this energy loss comes from bremsstrahlung. Electrons, one component of the plasma, have high thermal velocities. If these electrons collide with ions (nuclei) and are accelerated suddenly, electromagnetic waves are generated. Such waves are called bremsstrahlung. The energy of bremsstrahlung per unit volume and unit time is given by

$$P_B = 5.35 \times 10^{-37} n_e n_i T^{\frac{1}{2}} \text{ W/m}^3,$$
$$= 5.35 \times 10^{-37} n^2 T^{\frac{1}{2}} \text{ W/m}^3, \tag{1.38}$$

where $n_e$ is the electron number density and $n_i$ is the ion number density. In eq. (1.38), the temperature $T$ is expressed in keV. In the electrically-neutral plasma of mixed deuterium and tritium, $n_e = n_i = n$.

Let us now derive eq. (1.38). If an electron has an acceleration $\mathbf{w}$, the energy $S$ of the electromagnetic wave from the electron per unit time is

$$S = \frac{1}{6\pi\varepsilon} \frac{e^2}{c^3} w^2. \tag{1.39}$$

Here $-e$ is the electron charge and $c$ is the speed of light in vacuum. As $\mathbf{w}$ is a function of time $t$, the energy of the electromagnetic wave radiated by an electron is given by

$$\Delta E = \int_{-\infty}^{\infty} S \, dt = \frac{1}{6\pi\varepsilon} \frac{e^2}{c^3} \int_{-\infty}^{\infty} w^2 \, dt. \tag{1.40}$$

Consider the collision of an electron with an ion, with the collision (or impact) parameter $\rho_r$ and initial electron velocity $\mathbf{v}$, as shown in Fig. 1.5. The number of electrons in the velocity range between $\mathbf{v}$ and $\mathbf{v} + d\mathbf{v}$ is given by $f \, d\mathbf{v}$, where $f$ is the electron velocity distribution function. It is reasonable to suppose that the electron accelerations are the same for the collisions with initial electron velocities between $\mathbf{v}$ and $\mathbf{v} + d\mathbf{v}$ and with the collision parameters between $\rho_r$ and $\rho_r + d\rho_r$. The number of electrons which collide with one ion per unit time is $2\pi\rho_r v f\, d\rho_r\, d\mathbf{v}$. Since the energy of bremsstrahlung radiated from one electron collision is $\Delta E$, the energy of bremsstrahlung radiated by electrons described above is

$$q = 2\pi\rho_r v f \Delta E\, d\rho_r\, d\mathbf{v}. \tag{1.41}$$

In eq. (1.41), $d\mathbf{v} = dv_x\, dv_y\, dv_z$, where the suffixes $x$, $y$ and $z$ indicate the space components, respectively. When $f$ is symmetric with respect to $v_x, v_y$ and $v_z$, $f$ is a function of $v = |\mathbf{v}|$, and $f\, d\mathbf{v} = 4\pi v^2 f\, dv$. Then eq. (1.41) may be rewritten as

$$q = 8\pi^2 \rho_r v^3 f \Delta E\, d\rho_r\, dv. \tag{1.42}$$

The absolute value $w$ of the electron acceleration shown in Fig. 1.5 is estimated at

$$w \approx \frac{Ze^2}{4\pi\varepsilon\rho_r^2 m_e},$$

where $m_e$ is the electron mass, $Ze$ is the ion charge. The period $\Delta t$ during which the electron is accelerated is of the order of $\Delta t \approx \rho_r/v$. Thus we have

$$\Delta E \approx \frac{1}{6\pi\varepsilon}\frac{e^2}{c^3}w^2\Delta t \approx \frac{1}{(4\pi\varepsilon)^3}\frac{2}{3}\frac{Z^2 e^6}{m_e^2 c^3 \rho_r^3 v}.$$

On the other hand, the frequency $\nu$ of the radiated light accompanied by the acceleration during $\Delta t$ is related to $\Delta t$ by $\nu\Delta t \approx 1$. Thus given the relation

$$\rho_r \approx \frac{v}{\nu}, \qquad d\rho_r \approx -\frac{v}{\nu^2}\,d\nu \sim \frac{\rho_r^2}{v}\,d\nu,$$

**Fig. 1.5.** Coulomb collision between an electron and an ion.

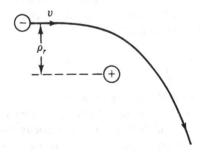

eq. (1.42) may be written as

$$q = 2\pi\rho_{\mathrm{r}} vf\Delta E \, \mathrm{d}v \, \mathrm{d}\rho_{\mathrm{r}}$$

$$\approx \frac{Z^2 e^6 v^2 f \, \mathrm{d}v}{12\pi\varepsilon^3 m_{\mathrm{e}}^2 c^3 \rho_{\mathrm{r}}^2} \, \mathrm{d}\rho_{\mathrm{r}} \approx \frac{Z^2 e^6 vf}{12\pi\varepsilon^3 m_{\mathrm{e}}^2 c^3} \, \mathrm{d}v \, \mathrm{d}v. \tag{1.43}$$

The electron distribution function $f$ is assumed here to have a Maxwellian form given by eq. (1.19). Equation (1.43) is integrated over the range from $v_{\min}$ to $\infty$ with regard to $v$, and from 0 to $\infty$ with regard to $v$. Substituting $f$ from eq. (1.19) in eq. (1.43) leads to

$$P_{\mathrm{B}} = \int_0^{\infty} \mathrm{d}v \int_{v_{\min}}^{\infty} \mathrm{d}v n_{\mathrm{t}} q = \int_0^{\infty} \mathrm{d}v \int_{v_{\min}}^{\infty} \mathrm{d}v \frac{Z^2 e^6 n_{\mathrm{t}} vf}{12\pi\varepsilon^3 m_{\mathrm{e}}^2 c^3}$$

$$= \int_0^{\infty} \mathrm{d}v \frac{1}{24\pi\varepsilon^3} \left(\frac{1}{2\pi m_{\mathrm{e}} T_{\mathrm{e}}}\right)^{\frac{1}{2}} \frac{Z^2 e^6}{m_{\mathrm{e}} c^3} n_{\mathrm{e}} n_{\mathrm{t}} \mathrm{e}^{-hv/kT}$$

$$= \frac{1}{24\pi\varepsilon^3} \left(\frac{kT_{\mathrm{e}}}{2\pi m_{\mathrm{e}}}\right)^{\frac{1}{2}} \frac{Z^2 e^6}{m_{\mathrm{e}} c^3 h} n_{\mathrm{e}} n_{\mathrm{t}}$$

$$= 5.35 \times 10^{-37} Z^2 n_{\mathrm{e}} n_{\mathrm{t}} T_{\mathrm{e}}^{\frac{1}{2}} \ \mathrm{W/m^3}, \tag{1.44}$$

where $h = 6.63 \times 10^{-34}$ Js is Planck's constant. Since the power $P_{\mathrm{B}}$ is given per unit volume, $P_{\mathrm{B}}$ in eq. (1.44) has been multiplied by the ion number density $n_{\mathrm{i}}$. In eq. (1.44), the lower limit $v_{\min}$ of the velocity has been chosen according to the relation

$$\tfrac{1}{2} m v_{\min}^2 = hv. \tag{1.45}$$

Note in eq. (1.44) that if the plasma of mixed deuterium and tritium has impurities of large $Z$, it will have a large energy loss, because the energy of bremsstrahlung is proportional to $Z^2$. In pure plasma which consists only of deuterium and tritium, $Z = 1$ and $n_{\mathrm{i}} = n_{\mathrm{e}} = n$.

### 1.3.3. Self-ignition temperature

The power $P_{\alpha}$ which is absorbed in the plasma by the self-heating of fusion reactions is given by eq. (1.37). As Fig. 1.6 indicates, $P_{\alpha}$ increases rapidly with temperature $T$ (strictly speaking, the ion temperature $T_{\mathrm{i}}$) in the region between 2 keV and 30 keV, according to the increase of $\langle\sigma v\rangle$. On the other hand, the power loss $P_{\mathrm{B}}$ by bremsstrahlung is proportional to $T^{\frac{1}{2}}$ (strictly speaking, to $T_{\mathrm{e}}^{\frac{1}{2}}$, where $T_{\mathrm{e}}$ is the electron temperature). At low temperatures, the plasma loses energy by bremsstrahlung, and has very little energy due to fusion reactions. If the number of fusion reactions increases, the self-heating of the plasma becomes significant; with the increase in plasma temperature, plasma heating outweighs energy loss by bremsstrahlung. The heating and the energy loss in the plasma of mixed deuterium and tritium is balanced at a temperature of $T_{si} = 4$ keV. The plasma maintains its high temperature, enabling it to continue the reactions, once it reaches this temperature. Thus the temperature $T_{si}$ is called the self-ignition temperature. As is clear from eq. (1.37), $P_\alpha$ is proportional to $n^2$. On the other hand, $P_B$ is also proportional to $n^2$, as can be seen from eq. (1.44). Therefore $T_{si}$ is independent of $n$.

### *1.3.4. The Lawson criterion*

In the preceding paragraph, the self-ignition temperature $T_{si}$ was obtained for the idealised (infinite) plasma. That value is too optimistic. The fusion plasma must be confined in a finite size. It is difficult to confine a plasma with a high temperature in a finite space, and then extract the fusion energy from the plasma. Due to the finite size of plasma, there are other energy losses in the plasma besides bremsstrahlung. Heat flux (including particle flux) occurs from the plasma to the surrounding wall. Energy loss by this heat flux per unit volume and per unit time is given by $3nkT/\tau$, where $\tau$ is the energy-confinement time. Since the thermal energy of particles per unit volume is given by $3nkT/2$, the sum of the thermal energies of electrons and ions is $3nkT$. The plasma is assumed to lose all thermal energy during $\tau$. If the energy loss by the heat flux is linear with the

**Fig. 1.6.** The power $P_\alpha$ of the fusion reaction absorbed in the plasma and the power loss $P_B$ by bremsstrahlung.

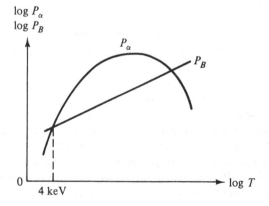

time, the energy equation for the plasma is

$$\frac{d3nkT}{dt} = P_\alpha - P_B - \frac{3nkT}{\tau} = \tfrac{1}{4}n^2\langle\sigma v\rangle E_\alpha - 5.35 \times 10^{-37}n^2T^{\frac{1}{2}} - \frac{3nkT}{\tau}.$$

(1.46)

The condition that the heat gains and losses are balanced in the plasma, so as to keep the temperature constant, can be written as

$$n\tau = \frac{12kT}{\langle\sigma v\rangle E_\alpha - 2.14 \times 10^{-36}T^{\frac{1}{2}}},$$

(1.47)

provided that the left-hand side of eq. (1.46) is zero. But eq. (1.47) seems too pessimistic: it may be modified to

$$\eta\left\{\tfrac{1}{4}\langle\sigma v\rangle n^2 E_f + P_B + \frac{3nkT}{\tau}\right\} = P_B + \frac{3nkT}{\tau}.$$

(1.48)

The right-hand side of eq. (1.48) gives the energy losses from the plasma. The first term in the bracket on the left-hand side is the total energy released by fusion reactions. The whole energy of bracketed terms is the sum of the fusion energy and the energy lost to the wall surrounding the plasma. When part of the energy is reflected into the plasma at a rate $\eta$, the plasma will maintain thermal

**Fig. 1.7.** The Lawson criterion.

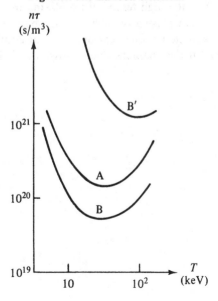

balance, giving eq. (1.48). Equation (1.48) reduces to

$$n\tau = \frac{24kT}{\langle\sigma v\rangle E_f - 4.28 \times 10^{-36}T^{\frac{1}{2}}}, \tag{1.49}$$

if $P_B$ in eq. (1.48) is replaced by the value of $P_B$ in eq. (1.44) and $\eta$ is taken to be $\frac{1}{3}$. The condition specified by eq. (1.49) is usually called the Lawson criterion. In Fig. 1.7, curve A shows $n\tau$ as given by eq. (1.47), and curve B shows $n\tau$ from eq. (1.49) for the D–T reaction. Curve B' shows $n\tau$ given by eq. (1.49) for the D–D reactions. Curve B indicates that $n\tau = 10^{20}$ s/m$^3$ at $T = 10$ keV. This value of $n\tau$ is about the minimum at which the fusion reaction continues in the plasma without an external energy supply. In a reactor, it is necessary that

$$n\tau > 10^{20} \text{ s/m}^3. \tag{1.50}$$

Condition (1.50) is called the Lawson criterion in the narrow sense, and is the condition for extracting fusion energy at $T = 10$ keV.

### 1.3.5. Confinement of plasma

The condition that must be met if the plasma is to generate energy through fusion reactions was derived in the preceding paragraph. This condition indicates that the plasma must maintain its high temperature and high density for some time. The first practical objective is thus to increase the plasma temperature to 10 keV. The second is to confine the high-temperature plasma in a finite space. No vessel can contain plasma of such high temperature. However, two methods, magnetic confinement and inertial confinement, are being developed. In the chapters that follow, we shall investigate these methods.

# 2

# The magnetic confinement
# method

•

*The plasma must be heated and confined in order to utilise the nuclear
fusion energy released from the plasma with a high temperature. There
is a method called magnetic confinement. Let us study several methods
of magnetic confinement including Tokomak, investigating motions of
charged particles in magnetic fields.*

## 2.1. The fundamentals of the magnetic confinement method

### 2.1.1. The motion of charged particles in magnetic fields

As noted before, there is no vessel which can confine plasma with
a temperature of the order of 100 million degrees. Vessels, even when
made of metals appropriate to high temperatures, can be used only
for temperatures less than several thousand degrees. Stars such as
the sun confine their giant plasma spheres by gravity. On earth, on
the other hand, new methods of confining plasmas must be devised
in place of gravity.

The plasma consists of charged particles. Charged particles cannot
cross magnetic field lines, but rotate around the lines. Therein is the
outline of the idea of a basket of magnetic field lines in which to
confine the plasma.

### Cyclotron motion

Let us investigate (Fig. 2.1) the motion of charged particles in
electromagnetic fields. Charged particles in motion induce
electromagnetic fields, which in turn affect the motion of the particles.
The analysis of plasma motions is not straightforward. In the case
where the number of charged particles is small, the electromagnetic
fields induced by the charged particles can be neglected in
comparison with fields applied from outside; collisions between
particles are also taken to be few in such a case. The equation of
motion for a single charged particle is

$$m \frac{dv}{dt} = q(\mathbf{E} + \mathbf{v} \times \mathbf{B}), \qquad (2.1)$$

where $m$ is the mass, $q$ is the charge and $\mathbf{v}$ is the velocity of the particle in an electric field $\mathbf{E}$ and magnetic flux density $\mathbf{B}$, respectively. (The magnetic flux density $\mathbf{B}$ is related to the magnetic field $\mathbf{H}$ through $\mathbf{B} = \mu\mathbf{H}$, where $\mu$ is the susceptibility (permeability) of a vacuum. Hereafter $\mathbf{B}$ is usually called the magnetic field, for simplicity.) In space, in the absence of $\mathbf{B}$, a charged particle moves with constant acceleration $q\mathbf{E}/m$. In space, in the absence of $\mathbf{E}$, a particle moves with an acceleration of $q\mathbf{v} \times \mathbf{B}/m$, which is perpendicular to $\mathbf{v}$. This acceleration does not change the magnitude of velocity but does induce curvature in its path. The magnetic field does not affect the particle by doing any work on it; the kinetic energy of the particle does not change. Let us assume that the magnetic field is constant in space and in time, and that the particle velocity is perpendicular to the magnetic field. The path of the particle is then a circle. Denote the radius of the circle by $\rho_r$ and the velocity component perpendicular to the magnetic field by $v_\perp$. Since the acceleration due to the magnetic field $qv_\perp \mathbf{B}/m$, the angular velocity $\omega_c$ and the frequency $v_c$ of the circular motion are

$$\omega_c = \frac{q\mathbf{B}}{m}, \qquad v_c = \frac{\omega_c}{2\pi} = \frac{q\mathbf{B}}{2\pi m} \qquad (2.2)$$

respectively. The value $v_c$ depends on $m$, $q$ and $\mathbf{B}$, and is called the cyclotron frequency, Larmor frequency or gyrofrequency. Equation

**Fig. 2.1.** Motion of a charged particle in a magnetic field.

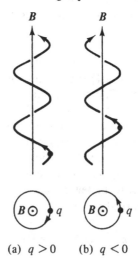

(a) $q > 0$     (b) $q < 0$

(2.2) reduces to

$$\omega_{ce} = 1.76 \times 10^{11} \mathbf{B} \text{ rad/s},$$

for an electron and to

$$\omega_{cp} = 0.96 \times 10^{8} \mathbf{B} \text{ rad/s},$$

for a proton. In these equations, $\mathbf{B}$ is expressed by Tesla. The centre of the circular motion is called the guiding centre. The radius $\rho_r$ of the circle

$$\rho_r = \frac{v}{\omega_c} = \frac{mv_\perp}{q\mathbf{B}} \tag{2.3}$$

is called the Larmor (gyro-)radius. In the case where the velocity $\mathbf{v}$ is not perpendicular to $\mathbf{B}$, the parallel component $v_{\parallel}$ of the velocity to the magnetic field is not changed by the magnetic field. Thus the particle motion is divided into two parts: a circular motion around the magnetic field, and a constant straight-line motion along the field. The result is that the particle executes spiral motion with a constant pitch. The particle rotates around a magnetic field line with a radius of $\rho_r$ and cannot cross this line. This fact forms the base for the magnetic confinement of the plasma. When a particle with charge $q$ executes circular motion with an angular velocity of $\omega_c$, the particle induces a circular current whose intensity is $I = \omega_c q/2$. Thus the circular motion of the charged particle may be considered to be a magnetic dipole $\mathbf{M}_m$. The magnitude $M_m$ is given by the product of $I$ and the area which the particle circles, that is,

$$\mathbf{M}_m = -I\pi\rho_r^2 \frac{\mathbf{B}}{B} = -\frac{\frac{1}{2} \cdot mv_\perp^2}{B^2}\mathbf{B} = -\frac{W_\perp}{B^2}\mathbf{B}. \tag{2.4}$$

Because the direction of the magnetic field induced by the circular motion of the charged particle is opposite to that applied from outside, the right-hand side of eq. (2.4) has a negative sign. In eq. (2.4), $W_\perp$ is the kinetic energy for the particle velocity component perpendicular to the magnetic field, $v_\perp$.

### Drift motion due to electric fields

Let us now consider the case where both $\mathbf{E}$ and $\mathbf{B}$ are constant in space and in time, and are perpendicular to each other. If we change the dependent variable in eq. (2.1) from $\mathbf{v}$ to $\mathbf{v}'$ through

$$\mathbf{v} = \mathbf{v}' + \frac{\mathbf{E} \times \mathbf{B}}{B^2}. \tag{2.5}$$

Then eq. (2.1) is transformed to

$$m\frac{d\mathbf{v'}}{dt} = q\left\{\mathbf{E} + \mathbf{v'} \times \mathbf{B} + \frac{1}{B^2}(\mathbf{E} \times \mathbf{B}) \times \mathbf{B}\right\}. \qquad (2.6)$$

$\mathbf{B} \cdot \mathbf{E} = 0$ because $\mathbf{B}$ is perpendicular to $\mathbf{E}$. Moreover,

$$(\mathbf{E} \times \mathbf{B}) \times \mathbf{B} = -B^2\mathbf{E},$$

so eq. (2.6) reduces to

$$m\frac{d\mathbf{v'}}{dt} = q\mathbf{v'} \times \mathbf{B}. \qquad (2.7)$$

Equation (2.7) gives circular motion with respect to $\mathbf{v'}$. Equation (2.8)

$$\mathbf{v}_D = \mathbf{E} \times \mathbf{B}/B^2, \qquad (2.8)$$

which is the second term on the right-hand side of eq. (2.5), shows a straight-line motion which is perpendicular to $\mathbf{E}$ and $\mathbf{B}$. The path of the particle is thus the sum of the circular and straight-line motions, describing a curve called a cycloid. This kind of particle motion, which crosses magnetic fields, is called drift motion. It is clear that drift motion plays an important role in the magnetic confinement of the plasma. If the component of $\mathbf{E}$ perpendicular to $\mathbf{B}$ is denoted by $E_\perp$ and the component parallel to $\mathbf{B}$ by $E_\parallel$, the drift velocity $v_D$ of the charged particle is given by

$$v_D = \frac{E_\perp}{B}\text{ m/s}, \qquad (2.7')$$

where $E_\perp$ is expressed in V/m. The direction and the magnitude of the drift velocity of the charged particle is constant regardless of its charge. The component $E_\parallel$ causes an acceleration of the particle along the magnetic field (Fig. 2.2).

**Fig. 2.2.** Drift motion due to an electric field.

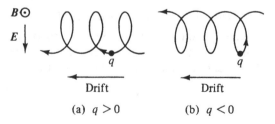

(a)  $q > 0$          (b)  $q < 0$

### Drift motion due to gravity

A charged particle undergoes drift motion when gravity has a component $mg_\perp$ perpendicular to the magnetic field. The field component of the gravity $mg_\perp/q$ is substituted for $E_\perp$ in eq. (2.7′), and the drift velocity $v_D$ is given by

$$v_D = \frac{mg_\perp}{qB} = \frac{g_\perp}{\omega_c}. \tag{2.8′}$$

The direction of the drift is dependent on the charge and is given by $g \times B$ for the positive charge (Fig. 2.3).

### Drift motion due to curvature of magnetic field

In the case in which the magnetic field has a curvature of radius $R$ in the absence of the electric and the gravitational field, a charged particle moving along the magnetic field with velocity $v_\parallel$ is affected by a centripetal acceleration of $v_\parallel^2/R$ instead of $g_\perp$, as was shown in eq. (2.8). Thus the drift velocity $v_D$ of the particle across the magnetic field is

$$v_D = \frac{mv_\parallel^2}{qBR}. \tag{2.9}$$

**Fig. 2.3.** Drift motion due to a gravitational field.

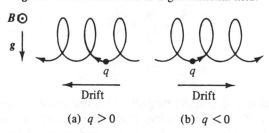

(a) $q > 0$       (b) $q < 0$

**Fig. 2.4.** Curvature of a magnetic field.

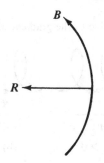

The drift motion is perpendicular to the magnetic field and to the curvature, and its direction depends on the charge (Fig. 2.4).

### *Drift motion due to the gradient of the magnetic field*

Although the direction of the magnetic field lines is constant, the magnitude of the field changes perpendicularly to the field, as is shown in Fig. 2.5. That is, **B** and $\nabla B$ are perpendicular to each other. According to eq. (2.3), the gyroradius is inversely proportional to $B$. The radius is small at a point where the magnetic field is strong, and the radius is large where the field is weak. Thus the particle undergoes a drift motion perpendicular to **B** and $\nabla B$, as is shown in Fig. 2.5. The drift velocity $v_D$ for this case cannot be obtained exactly, but is given approximately by

$$\frac{v_D}{v_\perp} = \frac{\rho_r|\nabla B|}{2B} \tag{2.10}$$

where $B/|\nabla B|$ is assumed to be smaller than $\rho_r$.

### *Motion due to divergence of the magnetic field*

Let us now study particle motion along the magnetic field. When the magnitude of the magnetic field changes in space, one of the four Maxwell equations for the magnetic field is

$$\text{div } B = \frac{\partial B_x}{\partial x} + \frac{\partial B_y}{\partial y} + \frac{\partial B_z}{\partial z}$$

$$= \frac{1}{r}\frac{\partial}{\partial r}(rB_r) + \frac{\partial B_\theta}{r\,\partial\theta} + \frac{\partial B_z}{\partial z} = 0. \tag{2.11}$$

As indicated in Fig. 2.6, the magnetic field line has a divergence in the $r$–$z$ plane if $\partial B_r/\partial r$ is positive. The term $(\partial/\partial r)(rB_r)$ in eq. (2.11) is integrated from 0 to $\rho_r$ with respect to $r$, and provided that $B_\theta = 0$,

**Fig. 2.5.** Drift motion due to the gradient of a magnetic field.

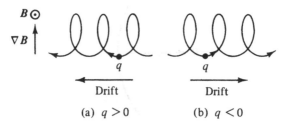

(a) $q > 0$          (b) $q < 0$

$\rho_r B_r$ is derived from

$$\rho_r B_r = -\int_0^{\rho_r} r \frac{\partial B_z}{\partial z} \, dr.$$

If the change in the magnetic field is assumed to be small, $\partial B_z/\partial z$ can be approximated by $\partial B/\partial z$. Thus we have

$$B_r = -\frac{\rho_r}{2} \frac{\partial B}{\partial z}.$$

Due to this $B_r$, a force $F_{\parallel}$ is given by

$$F_{\parallel} = q v_{\perp} B_r = -\frac{W_{\perp}}{B} \frac{\partial B}{\partial z} = (M_m \cdot \nabla) B,$$

and acts on the particle along the $z$-axis. The equation of motion along the field is

$$m \frac{dv_{\parallel}}{dt} = -\frac{W_{\perp}}{B} \frac{dB}{ds},$$

where $s$ is the distance along the field line. The above equation is rewritten as

$$m \frac{ds}{dt} \frac{dv_{\parallel}}{ds} = m v_{\parallel} \frac{dv_{\parallel}}{ds} = \frac{dW_{\parallel}}{ds} = -\frac{W_{\perp}}{B} \frac{dB}{ds}. \tag{2.12}$$

Since the magnetic field does not do any work on the particle, the total kinetic energy $W$ of the particle, which is the sum of the parallel component $W_{\parallel}$ and the perpendicular component $W_{\perp}$, is constant.

**Fig. 2.6.** Divergence of a magnetic field.

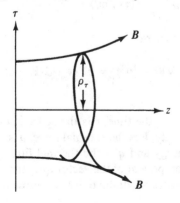

Thus we have $dW_\parallel/ds = -dW_\perp/ds$ and eq. (2.12) reduces to

$$\frac{dW_\perp}{ds} = \frac{W_\perp}{B}\frac{dB}{ds},$$

namely

$$\frac{d}{ds}\left(\frac{W_\perp}{B}\right) = 0. \qquad (2.13)$$

Equations (2.4) and (2.13) indicate that $M_m = W_\perp/B$ remains unchanged when the particle moves along a magnetic field whose intensity is not constant. The force $F_\parallel$ on the particle parallel to the field is given by

$$F_\parallel = -M_m\frac{dB}{ds} = -\frac{d}{dz}(M_m B) = -\nabla_\parallel \cdot (M_m B), \qquad (2.14)$$

in which $M_m B$ is taken to be the potential of the field.

### 2.1.2. Maxwell stress and pinch motion

Motion of a single particle in magnetic fields has been investigated above. In general, plasma consists of many charged particles. The plasma motions are frequently analysed by using the Navier–Stokes equation for a fluid with electrical conductivity. This method of analysis is called magnetohydrodynamics (MHD). The governing equations are the equation of continuity,

$$\frac{\partial \rho}{\partial t} + \frac{\partial \rho v_i}{\partial r_i} = 0, \qquad (2.15)$$

the equation of motion,

$$\rho\left\{\frac{\partial v_i}{\partial t} + (\mathbf{v}\cdot\nabla)v_i\right\} = \rho F_i - \frac{\partial p_{ij}}{\partial r_j}, \qquad (2.16)$$

and the equation of energy

$$\rho\left\{\frac{\partial e}{\partial t} + \frac{1}{2}\frac{\partial v^2}{\partial t} + (\mathbf{v}\cdot\nabla)(e + \tfrac{1}{2}v^2)\right\} = \rho\mathbf{F}\cdot\mathbf{v} + \mathbf{J}\cdot\mathbf{E} - \operatorname{div} q - \frac{\partial}{\partial r_j}(v_i p_{ij}).$$

$$(2.17)$$

In these equations, $t$ is the time, $r$ is the space coordinate, $\rho$ is the density, $\mathbf{v}$ is the velocity, $\mathbf{F}$ is the external force, $p$ is the stress tensor, $e$ is the internal energy and $q$ is the thermal flux. The suffix $i$ or $j$ refers to the $i$ or $j$ component of the vector quantity. Because plasma is a fluid with an electrical conductivity, the force from the external electromagnetic field acts on it. If an external electric field $\mathbf{E}$ and a magnetic field $\mathbf{B}$ are applied on the plasma, the electromagnetic force

$$\mathbf{F}' = \rho_e \mathbf{E} + \mathbf{J} \times \mathbf{B} \qquad (2.18)$$

acts beside the external force $\mathbf{F}$ on the plasma. In the above equations, $\rho_e$ is the electric charge density and $\mathbf{J}$ is the current density. $\mathbf{J} \times \mathbf{B}$ in eq. (2.18) is called the Lorentz force. On the other hand, the magnetic field is considered to induce the Maxwell stress in the space. The Maxwell stress

$$p'_{ij} = \frac{1}{\mu}(\tfrac{1}{2}B^2\delta_{ij} - B_iB_j) \qquad (2.19)$$

can be added to the stress of the fluid $p_{ij}$ instead of the Lorentz force. In eq. (2.19) $\mu = 4 \times 10^{-7}$ Henry/m is the susceptibility in vacuum. In the magnetohydrodynamics, new unknown variables $\mathbf{E}$, $\mathbf{B}$, $\mathbf{J}$, and $\rho_e$ are introduced in the fundamental equations (2.15)–(2.17), because the Navier–Stokes equations include the electromagnetic effects. Thus the Maxwell equations for the electromagnetic fields,

$$\mathrm{rot}\ \mathbf{E} + \frac{\partial \mathbf{B}}{\partial t} = 0, \qquad (2.20)$$

$$\mathrm{rot}\ \mathbf{H} - \frac{\partial \mathbf{D}}{\partial t} = \mathbf{J}, \qquad (2.21)$$

$$\mathrm{div}\ \mathbf{D} = \rho_e, \qquad (2.22)$$

$$\mathrm{div}\ \mathbf{B} = 0, \qquad (2.23)$$

and Ohm's law

$$\mathbf{J} = \rho_e\mathbf{v} + \sigma_e(\mathbf{E} + \mathbf{v} \times \mathbf{B}), \qquad (2.24)$$

must be combined, in order to solve these unknown variables simultaneously. In the above equations, $\sigma_e$ is the electrical conductivity of the plasma, $\mathbf{D}$ is the electric flux, and $\mathbf{H}$ is the magnetic field. They are related to the electric field $\mathbf{E}$ and the magnetic flux density $\mathbf{B}$ through

$$\mathbf{D} = \varepsilon\mathbf{E}, \qquad \mathbf{B} = \mu\mathbf{H}, \qquad (2.25)$$

respectively.

Magnetohydrodynamic problems can be solved by the Navier–Stokes equations for the plasma, together with the Maxwell equations for the electromagnetic fields. It is very difficult to solve these equations exactly, however. Usually they are approximated, taking care not to lose the essential physics in the process.

The dielectric constant in the vacuum is so small that $\varepsilon = 8.854 \times 10^{-12}$ Farad·m; the electric flux **D** is usually small. On the other hand, phenomena with high frequencies, such as light propagation, are not covered by MHD. Therefore the displacement current in eq. (2.21) is usually neglected. That is,

$$\frac{\partial \mathbf{D}}{\partial t} = 0. \tag{2.26}$$

Secondly, we can assume that the plasma has charge-neutrality, because the charge can move quickly due to the electrical conductivity. If we differentiate eq. (2.22) with respect to $t$ we have

$$\frac{\partial \rho_e}{\partial t} = \text{div}\, \frac{\partial \mathbf{D}}{\partial t}. \tag{2.27}$$

Hence the charge density remains constant over time, provided that the displacement current $\partial \mathbf{D}/\partial t$ is neglected. It turns out that the electrically conducting fluid is charge-neutral if the fluid is charge-neutral in the initial stage. In other words, at the same order of approximation, we can ignore the displacement current and the charge density. From this quasi-neutral (the charge is free but the electric field is taken into account) approximation, we omit the convection current $\rho_e \mathbf{v}$ in Ohm's law (2.24), which reduces to

$$\mathbf{J} = \sigma_e (\mathbf{E} + \mathbf{v} \times \mathbf{B}). \tag{2.28}$$

With (2.26), eq. (2.21) becomes

$$\text{rot}\, \mathbf{H} = \mathbf{J}. \tag{2.29}$$

If **J** from eq. (2.29) is substituted into (2.28), we have

$$\mathbf{E} = \frac{1}{\sigma_e} \text{rot}\, \frac{\mathbf{B}}{\mu} - \mathbf{v} \times \mathbf{B}. \tag{2.30}$$

By using the above three equations, **J** and **E** can be expressed by **B** and **v**. (**D** and **H** are replaced by **E** and **B** by eq. (2.25).) When **E** from eq. (2.30) is inserted in eq. (2.20), we get

$$\frac{\partial \mathbf{B}}{\partial t} = \text{rot}(\mathbf{v} \times \mathbf{B}) - \text{rot}\left(\frac{1}{\sigma_e} \text{rot}\, \frac{\mathbf{B}}{\mu}\right). \tag{2.31}$$

In MHD, one new variable **B** appears in the Navier–Stokes equations, and hence there is an additional equation in MHD, eq. (2.31), which is called the equation of induction. In the approximation to neglect the displacement current and charge density in the plasma, eq. (2.18) reduces to

$$\mathbf{F}' = -\mathbf{B} \times \text{rot} \frac{\mathbf{B}}{\mu}. \tag{2.32}$$

In plasma with a high temperature in which the electrical conductivity is taken to be infinite, eq. (2.31) becomes

$$\frac{\partial \mathbf{B}}{\partial t} = \text{rot}(\mathbf{v} \times \mathbf{B}). \tag{2.33}$$

This equation indicates that the change in the intensity of the magnetic field comes from the fluid velocity **v**. If we observe the phenomenon within the frame of the fluid, the magnetic field cannot be changed. In other words, 'the magnetic field is frozen with the fluid motion in the perfectly conducting fluid'. The perfectly conducting fluid cannot cross the magnetic field. This is the principle for magnetic confinement of the plasma from the point of view of MHD.

Let us consider a space in which the magnetic field is applied along the x-axis. Since $\mathbf{B} = (B, 0, 0)$, the Maxwell stress is expressed by

$$p' = \frac{B^2}{2\mu} \begin{bmatrix} -1 & 0 & 0 \\ 0 & 1 & 0 \\ 0 & 0 & 1 \end{bmatrix} = \frac{B^2}{2\mu} \begin{bmatrix} 1 & 0 & 0 \\ 0 & 1 & 0 \\ 0 & 0 & 1 \end{bmatrix} - \frac{B^2}{\mu} \begin{bmatrix} 1 & 0 & 0 \\ 0 & 0 & 0 \\ 0 & 0 & 0 \end{bmatrix}. \tag{2.34}$$

Equation (2.34) indicates that the magnetic field induces the isotropic pressure $B^2/2\mu$ on the electrically conductive fluid, in addition to the tension stress $B^2/\mu$ along the magnetic field.

**Fig. 2.7.** Pinch motion of a cylindrical plasma column.

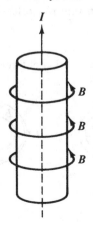

The magnetic field can be used to confine the plasma if we utilise the fact that the magnetic field puts a pressure on the plasma perpendicular to the magnetic field. When an electric current flows along the axis in the cylindrical plasma, as shown in Fig. 2.7, the magnetic field is induced in the azimuthal direction; this magnetic field exerts a pressure on the plasma and increases its temperature and the density. This phenomenon is called the pinch effect. Furthermore, the electric current increases the temperature of the plasma by the Joule heating. If the plasma is a straight cylinder, it has two ends which are in contact with the walls. For the torus plasma (Fig. 2.8), (a) the azimuthal magnetic field due to the axial current pinches the plasma ($z$-pinch) and (b) the axial magnetic field due to the azimuthal current pinches the plasma ($\theta$-pinch).

### 2.1.3. Sausage-type instability

Although the magnetic field can be used to confine plasma with high temperature and high density through the pinch effect of the Maxwell stress perpendicular to the magnetic field, the magnetic field induces the instability in the plasma column through the Maxwell stress itself. Let us suppose that the plasma column has a constricted part. The Stokes integral theorem is applied to a Maxwell equation rot $\mathbf{B}/\mu = \mathbf{j}$ to obtain

$$\oint \frac{\mathbf{B}}{\mu} \cdot d\mathbf{l} = \int\int j_n \, dS = I. \tag{2.35}$$

Here $S$ is chosen as the cross-sectional area of the plasma column and $j_n$ is the component of the current density normal to the cross-section. The total current across the cross-section is denoted here by $I$ and the circumference vector of the section $S$ by $\mathbf{l}$. The total current $I$ remains constant regardless of the cross-section even

Fig. 2.8. Pinch motion of toroidal plasma.

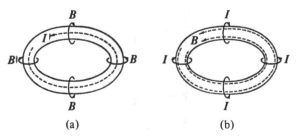

(a)                    (b)

if the plasma has a constricted part shown in Fig. 2.9. If eq. (2.35) is applied to the cross-section where the radius is $r$, we have

$$2\pi r \frac{B_\theta}{\mu} = I, \qquad (2.36)$$

which shows that $B_\theta$ is inversely proportional to $r$, where the suffix $\theta$ refers to the component in the azimuthal direction. It turns out from eq. (2.36) that a cross-section with smaller $r$ shrinks more through the increase of the Maxwell stress in the magnetic field. The constriction grows to form a neck and at last the current is shut down with the cutting off of the plasma column. The plasma column inside which the electric current flows along the axis is unstable. Such an instability is called a 'sausage-type' instability because of its form.

By applying the magnetic field $B_z$ along the axis of the plasma column, the sausage-type instability can be stabilised. The plasma cannot move across the magnetic field because plasma with high temperature has good electrical conductivity. Accordingly, the total magnetic flux across the cross-section remains constant, even if the cross-sectional area changes. That is, the equality

$$\pi r^2 B_z = \text{const.} \qquad (2.37)$$

obtains between $r$ and $B_z$. Equation (2.37) expresses that $B_z$ is inversely proportional to $r^2$. The increasing $B_z$ with the decreasing $r$ increases the Maxwell stress in the radial direction (perpendicular to the axis). This increase in the magnetic pressure affects the plasma

**Fig. 2.9.** Sausage-type instability.

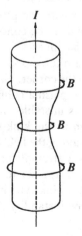

column, expanding the radius of the shrinking cross-section. Differentiation of the logarithm of both sides of eqs (2.36) and (2.37) gives

$$\frac{\delta B_\theta}{B_\theta} + \frac{\delta r}{r} = 0, \qquad (2.38)$$

$$\frac{\delta B_z}{B_z} + \frac{2\delta r}{r} = 0. \qquad (2.39)$$

As described above, $B_\theta$ causes the section to shrink while $B_z$ causes it to expand (if we neglect the change of $B_z$ outside the plasma column). The outward pressure component in the radial direction of the Maxwell stress is given by $p' = B_z^2/2\mu - B_\theta^2/2\mu$. The plasma column is stable when $\delta p'$ is positive for $-\delta r > 0$ (decrease in the radius). Equations (2.38) and (2.39) combine to give

$$\delta p' = \delta\left(\frac{B_z^2}{2\mu} - \frac{B_\theta^2}{2\mu}\right) = \frac{B_z\delta B_z}{\mu} - \frac{B_\theta\delta B_\theta}{\mu} = \frac{2B_z^2\delta r}{\mu r} + \frac{B_\theta^2\delta r}{\mu r}$$

$$= \frac{1}{\mu r}(2B_z^2 - B_\theta^2)(-\delta r).$$

Thus we have

$$2B_z^2 > B_\theta^2, \qquad (2.40)$$

as the stability condition of the plasma column. When a strong component $B_z$ of the magnetic field is applied in advance to the plasma column, the sausage-type instability is stabilised.

### 2.1.4. Kink-type instability

Next let us study the case where the plasma column bends, as in Fig. 2.10. The magnetic field lines are concentrated and the magnetic flux density increases at the inner side of the corner, while the magnetic field lines become diluted and the magnetic flux density decreases at the outer side of the corner. The difference in the pressure of the Maxwell stress on the plasma column increases the bending angle. The plasma column is unstable under this kind of bending, and the instability is known as the kink-type instability. Since the magnetic field has a tension along the field line, $B_z$ acts on the plasma to stretch the column into a straight line, provided the magnetic field in the axial direction is applied to the plasma column. The stabilising effect of $B_z$ on the kink-type instability, however, has a limit, unlike the sausage-type instability (Fig. 2.11).

Here let us imagine that the plasma column of height $2\lambda$ bends and has a curvature $R$. Two end plates A and A' have an inclination angle $\alpha = \lambda/R$ to the horizontal plane and lie in the planes $S$ and $S'$, respectively. There is a magnetic field $B_\theta$ in the plane $S$. If the total current in the plasma column is denoted by $I$, $B_\theta$ at the radius $r$ in the plane $S$ is given by $B_\theta = \mu I/2\pi r$ from eq. (2.36). This $B_\theta$ induces pressure $p'$ perpendicular to the plane $S$ by the Maxwell stress: $p' = B_\theta^2/2\mu$. The resultant force $F$ perpendicular to the plane $S$ is written as

$$F = \int_a^\infty p' 2\pi r \, dr = \int_a^\infty \frac{B_\theta^2}{2\mu} 2\pi r \, dr = \int_a^\infty \frac{\mu I^2}{4\pi r} \, dr,$$

where $a$ is the radius of the plasma column. In the above equation,

**Fig. 2.10.** Kink-type instability.

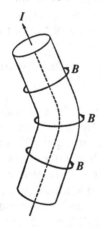

**Fig. 2.11.** Kink-type instability and magnetic field.

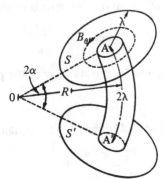

the Maxwell stress on the area A is neglected. If we consider the fact that the plasma column of a height of only $2\lambda$ has a curvature, the Maxwell stress on the plane $S$ is effective in the region whose radius $r$ is smaller than $\lambda$. Therefore the integration with respect to $r$ is carried out in the region from $a$ to $\lambda$ instead of the region from $a$ to $\infty$. Thus

$$F = \int_a^\lambda \frac{\mu I^2}{4\pi r}\, dr = \frac{\mu I^2}{4\pi} \log \frac{\lambda}{a}.$$

Since we have $B_\theta(a) = \mu I/2\pi a$, if we denote $B_\theta$ at $r = a$ by $B_\theta(a)$, the above equation becomes

$$F = \frac{\pi a^2 B_\theta^2(a)}{\mu} \log \frac{\lambda}{a}. \tag{2.41}$$

On the bending part of the plasma column, two forces of the amplitude of $F$ act perpendicular to the planes of $S$ and $S'$. The planes $S$ and $S'$ have an inclination angle $\alpha$ to the horizontal plane, so the resultant component $F_R$ of the two forces in the horizontal direction is given by

$$F_R = 2F \sin \alpha \approx \frac{2\pi \lambda a^2 B_\theta^2(a)}{\mu R} \log \frac{\lambda}{a}. \tag{2.42}$$

The Maxwell stress acting on the side wall of the plasma column along the height of $2\lambda$ is stronger at the inner part of the bending angle and is weaker at the outer part. This stress also increases the bending angle. But the resultant force of this Maxwell stress is smaller than $F_R$, and we may disregard its effect here. On the other hand, the component $B_z$ of the magnetic field (Fig. 2.12) induces the tension $H = \pi a^2 B_z^2/\mu$ at sections A and A', provided that $B_z$ is uniform through the cross-sectional area of the plasma column. The component $H_{-R}$ of the resultant force in the horizontal direction (to restore the straightness of the column) is expressed by

$$H_{-R} = 2H \sin \alpha \approx \frac{2\pi \lambda a^2 B_z^2}{\mu R}. \tag{2.43}$$

The condition required for the stability of the plasma column is given by

$$H_{-R} > F_R.$$

By using eqs (2.42) and (2.43), the above inequality becomes

$$\frac{B_z^2}{\mu_p} > \frac{B_\theta^2(a)}{\mu} \log \frac{\lambda}{a}. \tag{2.44}$$

In the state of equilibrium, the acceleration of the plasma is zero. Thus the left-hand side of the equation of motion (2.16) is zero. The forces on the right-hand side of eq. (2.16) are in balance with each other. The stress acting on the plasma consists of the pressure $p$ and the Maxwell stress $p'$. On the circular surface of the plasma column, the outward pressure inside the plasma is given by $p + B_z^2/2\mu$, while the inward pressure outside the plasma is given by $B_\theta^2/2\mu$. (It is assumed that $B_\theta = 0$ inside the plasma and $B_z = 0$ outside the plasma). In the state of equilibrium, both the pressures are equal. Thus we have

$$\frac{B_\theta^2(a)}{2\mu} = p + \frac{B_z^2}{2\mu}. \tag{2.45}$$

Equation (2.45) requires that $B_\theta^2(a)/2\mu > B_z^2/2\mu$. Given this

Fig. 2.12. Forces acting in bending the plasma column.

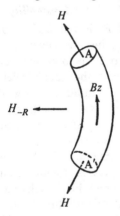

Fig. 2.13. Plasma column surrounded by a metal shell (liner).

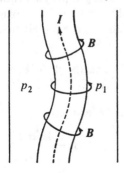

condition, inequality (2.44) is not satisfied in the case where $\lambda$ is much longer than $a$. In other words, kink-type instability with a long wavelength of bending is not stabilised by the magnetic field in the axial direction.

In order to stabilise the kink-type instability, the plasma column is covered by a metal shell, as shown in Fig. 2.13. The magnetic field cannot cross the metal surface, because the metal has a large electrical conductivity. When the plasma column bends in the metal shell, as in Fig. 2.13, $p_1$ is less than $p_2$. The magnetic field lines at $p_1$ push back the plasma column, against the bending. Thus the metal shell stabilises the kink-type instability of the plasma column. According to this analysis, the plasma column becomes stable for the kink-type instability when the radius of the metal shell is less than five times the plasma radius. The metal shell stabilising the kink-type instability is frequently called a 'liner'.

### 2.1.5. *Spiral instability*

A strong magnetic field $B_z$ is required in order to stabilise the sausage-type instability as just described. There is another reason why a strong $B_z$ is required. If the total current in the plasma column along the axis is denoted by $I$, the magnetic field $B_\theta(a)$ on the surface of the plasma column whose radius is $a$, is given by $B_\theta(a) = \mu I/2\pi a$. The magnetic field line on the surface of the plasma column describes a spiral in the case where $B_\theta(a)$ is combined with the externally applied magnetic field $B_z$. If a perturbation of the spiral appears on the surface of the plasma column, with the same pitch as that of the magnetic field line, the plasma column is unstable because the magnetic field cannot suppress the growth of a perturbation parallel to the field. This kind of perturbation is called the spiral-type or screw-type instability. The spiral-type instability grows when the pitch $\lambda$ is shorter than the length (height) $L$ of the plasma column. In the case where $L$ is shorter than $\lambda$, the plasma column is stable for the spiral-type instability. Since the pitch $\lambda$ of the magnetic field line is expressed by

$$\lambda = \frac{2\pi a B_z}{B_\theta(a)} = \frac{4\pi^2 a^2 B_z}{\mu I}, \tag{2.46}$$

the stability condition $L < \lambda$ leads to

$$I < I_c, \qquad I_c = \frac{4\pi^2 a^2 B_z}{\mu L}. \tag{2.47}$$

In order that the plasma column can be held stable, the current in the plasma column has the upper limit $I_c$. This $I_c$ is called the Kruskal–Shafranov limit.

### 2.1.6. The β-value

As described above, the equation

$$p(r) + \frac{B^2(r)}{2\mu} = \frac{B^2(a)}{2\mu} \qquad (2.48)$$

must obtain for plasma in equilibrium. $B(a)$ is the magnetic field on the surface of a plasma column whose radius is $r = a$, and is the vector sum of $\mathbf{B}_z(a)$ and $\mathbf{B}_\theta(a)$. The ratio of the plasma pressure $p(0)$ at the centre to the magnetic pressure $B(a)^2/2\mu$ on the surface of the plasma column

$$\beta = \frac{2\mu p(0)}{B^2(a)} \qquad (2.49)$$

is called the β-value. When the β-value is high, a weak magnetic field confines the plasma at high pressure. But $B_z$ must be strong in comparison with $B_\theta(a)$, in order to give a stable plasma column for either the sausage-type or the spiral-type instability, or both together. In the case where $B_z$ is constant inside and outside the plasma column, $p(0)$ must be equal to $B_\theta^2(a)/2\mu$. Therefore, the β-value is related by $\beta = B_\theta^2(a)/[B_z^2 + B_\theta^2(a)]$. That is, the β-value is small and hence $p(0)$ is also small. The plasma pressure $p$ is related to the number density $n$ and the temperature $T$ through $p = nk(T_e + T_i)$, where $k$ is the Boltzmann constant and the suffices e and i refer to the electron and the ion, respectively. In the same way, the internal energy $e$ of the plasma is expressed by $e = 3nk(T_e + T_i)/2$. The energy of the magnetic field per unit volume, on the other hand, is given by $B^2/2\mu$. Thus the β-value indicates the ratio of the plasma energy to the energy of the magnetic field by which plasma is confined, and so it is a measure of the economy of the device to confine the plasma.

## 2.2. The Tokomak

### 2.2.1. Structure of the Tokomak

A representative device for magnetic confinement is the Tokomak, which utilises the principles described in the preceding section. To remove the end sections of the plasma column, the Tokomak uses a form of the torus. The radius of the torus is called the major radius, the radius of the plasma column is called the minor radius. A transformer is used to induce the electric current into the plasma along the central axis (sub-axis) of the plasma column. The iron yoke of the transformer is set perpendicular to the torus at the centre of the major radius. The torus plasma plays the role of the secondary coil of the transformer. When the electric current flows in the primary coil, the magnetic flux in the iron yoke changes and the current is induced along the sub-axis in the toroidal plasma as the secondary coil. This induced current (toroidal current) generates the magnetic field $B_\theta$ around the plasma column to pinch it. $B_\theta$ is called the poloidal field. In order to prevent the plasma from generating either the sausage-type or the spiral-type instability, or both together, a magnetic field $B_z$ in the direction of the sub-axis has to be applied. This magnetic field is generated by coils wound around the plasma column along the azimuthal direction. $B_z$ is called the toroidal field. To stabilise kink-type instability, the toroidal discharge tube is made of metal. An outline drawing of the Tokomak is shown in Fig. 2.14.

The current $I$ in the torus plasma is approximately constant in the Tokomak independently of the plasma impedance, because the impedance of the circuit system is much larger than the plasma impedance itself. If we assume that the plasma is uniform in cross-sectional area, the Joule heating $W$ of the plasma per unit volume and unit time is written by

$$W = \frac{J^2}{c_e} = \frac{1}{c_e}\,(I/\pi a^2). \qquad (2.50)$$

**Fig. 2.14.** The Tokomak.

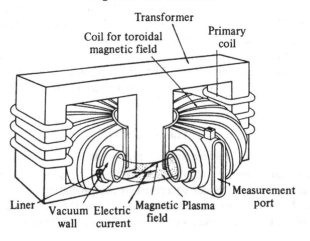

Transformer
Coil for toroidal magnetic field
Primary coil
Measurement port
Liner   Vacuum wall   Electric current   Magnetic field   Plasma

Here $a$ is the minor radius, $c_e$ is the electrical conductivity of the plasma and $J$ is the current density. The plasma temperature increases by Joule heating.

The aim of the recently built large Tokomak is to realise the critical condition (the lowest condition to satisfy the Lawson criterion) and to eliminate the iron yoke of the transformer, because the vicinity of the principal axis is crowded by toroidal coils and other parts. Only the primary coil is set inside the toroidal plasma around the principal axis to induce the toroidal current in the plasma by changing the magnetic flux in the vacuum along the principal axis. At present, the Tokomak is one of the most advanced among the devices for magnetic confinement fusion. In the Tokomak the current along the torus plasma plays two important roles: Joule heating, and the confinement of plasma by the induced poloidal magnetic field. The confined plasma is generally stable because the poloidal magnetic field is a minimum at the minor axis in the toroidal plasma. Due to the electric current flowing in the plasma, however, there are cases in which the plasma is unstable in comparison with the plasma in a device with the external coil system (stellarator or heliotron), which will be described later. The disruptive instability occurring in the Tokomak plasma sometimes damages the device. One of the fundamental problems with the Tokomak, which uses a transformer to obtain the toroidal current in the plasma, is to find how the toroidal current can be sustained in order to operate the device steadily.

### 2.2.2. Electrical conductivity of plasma

Let us now look at the electric conductivity of plasma. The conductivity of plasma depends on collisions between electrons and ions. An ion has a large mass in comparison with an electron. Generally speaking, the velocities of ions as well as their changes in momenta and energies in collisions are small in comparison with those of electrons. Therefore we may consider ions to be at rest in collisions with electrons. An electron which moves with the velocity **v** towards an ion moves along a hyperbolic line due to the Coulomb force between the ion and the electron, as is shown in Fig. 2.15. In the figure, $\rho_r$ is the impact parameter and $\chi$ is the angle of deflection. For the Coulomb force, angle $\chi$, which is defined by

$$\chi = \pi - 2\varphi, \tag{2.51}$$

satisfies

$$\tan \varphi = 4\pi\varepsilon \frac{m_e \rho_r v^2}{Ze^2}, \tag{2.52}$$

where $\varepsilon$ is the dielectric constant of a vacuum, $m_e$ is the electron mass and $Ze$ is the ion charge, respectively. Let us consider a case in which the trace of an electron turns 90° after a collision. The suffix 0 for the impact parameter refers to $\tan \varphi = 1$ in eq. (2.52). That is,

$$\rho_{r0} = \frac{Ze^2}{4\pi\varepsilon m_e v^2}. \tag{2.53}$$

Because the Coulomb force is inversely proportional to the square of the distance, the electron is reflected only a little for a large $\rho_r$. This is not the only way to define 'collision' using the deflection angle $\chi$; here a collision is defined by deflection angles greater than 90°. Then the collision cross-sectional area $\sigma_0$ is given by

$$\sigma_0 = \pi\rho_{r0}^2 = \frac{Z^2 e^4}{16\pi\varepsilon^2 m_e^2 v^4}. \tag{2.54}$$

The mean free path $l_0$ is derived as

$$l_0 = \frac{1}{n\sigma_0} = \frac{16\pi\varepsilon^2 m_e^2 v^4}{Z^2 e^4 n}, \tag{2.55}$$

by using eq. (1.22). In this equation, $n$ is the number density of the electron. The mean free path changes significantly with the electron velocity, since it depends on $v^4$. If the velocity $v$ is represented by the thermal velocity $\bar{v} = 3kT/m_e$ defined by (1.18), eq. (2.55) becomes

$$l_0 = \frac{144\pi\varepsilon^2 m_e^2 k^2 T^2}{Z^2 e^4 n}. \tag{2.56}$$

Equation (2.56) shows that $l_0$ is proportional to $T^2$. It is easy to see that the mean free path becomes long and collisions between particles become scarce when the temperature of the plasma is high.

The angle of deflection is small when $\rho_r > \rho_{r0}$. But a sequence of small angles of deflection gives the same contribution to the

**Fig. 2.15.** Collision between an electron and an ion.

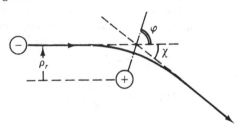

phenomena as one large angle of deflection. This accumulation of small scattering angles is called multiple scattering. Equation (2.52) is approximated by

$$\sin^2 \chi = 4\left(\frac{\rho_{r0}}{\rho_r}\right)^2\left\{1+\left(\frac{\rho_{r0}}{\rho_r}\right)^2\right\}^{-2},$$

or by the simpler form

$$\chi^2 = 4\left(\frac{\rho_{r0}}{\rho_r}\right)^2,$$

for $\rho_r > \rho_{r0}$. The number of collisions per unit time with impact parameters between $\rho_r$ and $\rho_r + d\rho_r$ is expressed by $nv2\pi\rho_r\, d\rho_r$. Thus the mean value $\langle\chi^2\rangle$ is given by

$$\langle\chi^2\rangle = nv2\pi \int_{\rho_{min}}^{\rho_{max}} \chi^2\rho_r\, d\rho_r = 8\pi\rho_{r0}^2 nv \log\frac{\rho_{rmax}}{\rho_{rmin}}.$$

Since $\langle\chi^2\rangle$ in the above equation is the mean for small scattering angles only, we choose $\rho_{rmin} = \rho_{r0}$ and $\rho_{rmax} = r_D$. Here $r_D$ is the Debye radius and is expressed by

$$r_D = \left(\frac{\varepsilon kT_e}{ne^2}\right)^{\frac{1}{2}} = 6.90 \times 10^3 \left(\frac{T_e}{n}\right)^{\frac{1}{2}} \text{ m}. \qquad (2.57)$$

The Debye radius is the distance over which a charged particle in practice influences its electric field.

The Debye radius is derived in the following way. An ion is located at one point in space. It is expected that there are more electrons than ions around this ion. Our task is to derive the equilibrium electron distribution around this ion. The electric potential $\varphi$ in the neighbourhood of the ion depends on the charge of the ion and the electron cloud around the ion. The mean electron number density is denoted by $n$. The electron number density $n_e$ around the ion deviates a little from $n$. That is,

$$n_e(r) \neq n.$$

The equilibrium distribution at a point where the potential is $\varphi$ is given by

$$n_e(r) = n \exp\left(\frac{e\varphi}{kT_e}\right),$$

where $T_e$ is the electron temperature. Since the perturbed electric potential is considered to be small, the above equation may be expanded in $\varphi$ to give

$$n_e(r) = n \exp\left(1 + \frac{e\varphi}{kT_e}\right). \tag{2.58}$$

On the other hand, the Laplace equation for $\varphi$ is given by

$$\varepsilon\Delta\varphi = (n_e - n)e, \tag{2.59}$$

in which $n_e$ is eliminated by using eq. (2.58) to yield

$$\Delta\varphi = \frac{ne^2\varphi}{\varepsilon kT_e}. \tag{2.60}$$

If we denote $r_D$ by

$$r_D = \left(\frac{\varepsilon kT_e}{ne^2}\right)^{\frac{1}{2}}, \tag{2.61}$$

and take into account the spherical symmetry of the phenomenon, eq. (2.60) may be rewritten as

$$\frac{1}{r^2}\frac{d}{dr}\left(r^2\frac{d\varphi}{dr}\right) = \frac{\varphi}{r_D^2}. \tag{2.62}$$

Among the infinite series of solutions of this second-order differential equation, we choose a solution which satisfies $\varphi \to 0$ as $r \to \infty$ and $\varphi \to e/(4\pi\varepsilon r)$ as $r \to 0$. Thus we obtain

$$\varphi = \frac{e}{4\pi\varepsilon r}\exp\left(-\frac{r}{r_D}\right). \tag{2.63}$$

In the above equation, $r_D$ is the Debye radius. The potential around a charged particle becomes the Yukawa type instead of the Coulomb type, because of the shielding effect induced by the concentration of the particles with opposite charge around a charged particle. The potential $\varphi$ given by eq. (2.63) decays faster than the Coulomb potential as $r$ increases. The distance over which the potential is effective if $r_D > r$.

The accumulation of small scattering angles contributes in the same order of magnitude to the phenomenon as does a large scattering angle provided an electron moves during the time period of $\tau_m$ satisfied by $\langle\chi^2\rangle\tau_m \approx 1$. The mean free path $l_m$ of the multiple scattering is expressed by $l_m = v\tau_m$. The effective cross-sectional area $\sigma_m$ of the multiple scattering is obtained through $l_m = 1/(n\sigma_m)$ as

$$\sigma_m = \frac{1}{nl_m} = \frac{1}{nv\tau_m} = \frac{\langle\chi^2\rangle}{nv} = 8\pi\rho_{r0}^2 \log\frac{r_D}{\rho_{r0}}. \tag{2.64}$$

Table 2.1 *The Coulomb logarithm* ($\log \Lambda$)

| T (K) \ $n_e$ (m$^{-3}$) | $10^6$ | $10^9$ | $10^{12}$ | $10^{15}$ | $10^{18}$ | $10^{21}$ | $10^{24}$ | $10^{21}$ | $10^{30}$ |
|---|---|---|---|---|---|---|---|---|---|
| $10^2$ | 16.3 | 12.8 | 9.43 | 5.97 | | | | | |
| $10^3$ | 19.2 | 16.3 | 12.8 | 9.43 | 5.97 | | | | |
| $10^4$ | 23.2 | 19.7 | 16.3 | 12.8 | 9.43 | 5.97 | | | |
| $10^5$ | 26.7 | 23.2 | 19.7 | 16.3 | 12.8 | 9.43 | 5.97 | | |
| $10^6$ | 29.7 | 26.3 | 22.8 | 19.3 | 15.9 | 12.4 | 8.96 | 5.54 | |
| $10^7$ | 32.0 | 28.5 | 25.1 | 21.6 | 18.1 | 14.7 | 11.2 | 7.85 | 4.39 |
| $10^8$ | 34.3 | 30.9 | 27.4 | 24.0 | 20.5 | 17.0 | 13.6 | 10.1 | 6.69 |

As is clear from Table 2.1, generally

$$\frac{\sigma_m}{\sigma_0} = 8 \log \frac{r_D}{\rho_{r0}} \gg 1,$$

and the multiple scattering contributes a large effect to the phenomena related to the collision.

Thus $\sigma_m$ may be used as the effective cross-sectional area for the collision between the electron and the ion. The mean collision time $\tau'$, which is the mean period during which an electron moves without colliding with an ion, is really the same as $\tau_m$ and is given by

$$\tau' = \tau_m = \frac{l_m}{v} = \frac{1}{\sigma_m n v} = \frac{1}{8\pi \rho_{r0}^2 n v \log r_D/\rho_{r0}}.$$

If $\rho_{r0}$ of eq. (2.53) is put into the above equation, and $v$ is represented by $\bar{v}$ and expressed in $T_e$ by $\frac{1}{2}m\bar{v}^2 = \frac{3}{2}kT_e$, then the mean collision time $\tau$ is given by

$$\tau = \frac{2\pi\varepsilon^2 m_e^{\frac{1}{2}}(3kT_e)^{\frac{3}{2}}}{nZ^2 e^4 \log \Lambda}, \qquad (2.65)$$

where $\log \Lambda$ is

$$\log \Lambda = \log \frac{3\varepsilon}{Ze^3}\left(\frac{4\pi k^3 T_e^3}{n}\right)^{\frac{1}{2}}, \qquad (2.66)$$

and is called the Coulomb logarithm. The values (for $Z = 1$) are shown in Table 2.1. When an external electric field $E$ is applied to the plasma, the distance $d$ through which an electron with the initial velocity of zero passes during the mean collision time $\tau$ is given by

$$d = \frac{eE}{2m_e}\tau^2.$$

Just after a collision of an electron with an ion, the mean velocity along the electric field is taken to be zero. Thus the mean velocity $v_E$ of the electron along the electric field is

$$v_E = \frac{d}{\tau} = \frac{eE}{2m_e^2}\,\tau.$$

The current density $j$ is

$$J = nev_E = \frac{ne^2}{2m_e}\,\tau E.$$

If $\tau$ in eq. (2.65) is substituted in the above equation, we have

$$J = \frac{\pi\varepsilon^2 (3kT_e)^{\frac{3}{2}}}{Z^2 e^2 m_e^{\frac{1}{2}} \log \Lambda}\,E.$$

On the other hand, the current density $J$ is related to the electric field $E$ and the electric conductivity $c_e$ by

$$J = c_e E.$$

Thus we have

$$c_e = \frac{\pi\varepsilon^2 (3kT_e)^{\frac{3}{2}}}{Ze^2 m_e^{\frac{1}{2}} \log \Lambda}. \tag{2.67}$$

In eq. (2.57) or eq. (2.61), the charge number of the ion is taken to be 1. In eq. (2.67), the charge number is denoted by $Z$. If $T_e$ is expresssed in eV, then eq. (2.67) becomes

$$c_e = 5.2\,\frac{T_e^{\frac{3}{2}}}{Z \log \Lambda}\,\text{mho/m}. \tag{2.68}$$

As is described above, the total current $I$ in the Tokomak plasma is constant, independent of the electric conductivity of the plasma. The Joule heating per unit volume and unit time is inversely proportional to $c_e$ as eq. (2.50) shows. Thus the Joule heating is inversely proportional to $T_e^{\frac{3}{2}}$, as eq. (2.50) shows. Plasma with a high temperature is not heated any more by the electric current. In the Tokomak, in which an electric current flows along the sub-axis, the electron temperature increases up to $10^7$ K (1 keV) by the Joule heating. It is necessary to utilise supplementary heating methods to raise the plasma temperature one order of magnitude.

### 2.2.3. *Neutral-beam heating*

The Joule heating for the Tokomak is called the primary heating. Heating to increase the plasma temperature from 1 keV to 10 keV after the Joule heating is called the secondary heating. One of the most important secondary heating methods is neutral-beam injection into the plasma.

If ions are injected into the plasma at high speed from outside the Tokomak, the ion temperature in the plasma increases by colliding with high-speed ions. Toroidal and poloidal magnetic fields with strong intensities confine the torus plasma in the Tokomak; these magnetic fields prevent the charged particles from escaping from the plasma. But it is therefore impossible to inject charged particles into the torus plasma from the outside across the magnetic fields. So a high-speed neutral beam is used instead.

Electrically neutral particles can be injected into plasma, for they are neither interrupted by the magnetic fields around the plasma, nor do they disturb the magnetic fields which confine the plasma. The high-speed neutral particles collide with ions in the plasma. The radius of the electron orbit around the nucleus of the neutral particle is $10^5$ times greater than the radius of the nucleus itself. Thus the orbit electrons of the neutral particles interact with ions in the plasma during collisions. As a result, high-speed neutral particles become high-speed ions, being deprived of their electrons by ions in the plasma, and ions in the plasma become low-speed neutral particles on acquiring electrons. Thus by using these high-speed neutral beams, the plasma temperature can be increased to the corresponding value of the speed of the injected neutral beams.

A method called parallel injection, by which neutral beams are injected parallel to the toroidal current $I$ in the Tokomak, is designated by A in Fig. 2.16. There is another method, called

**Fig. 2.16.** The secondary heating of the plasma by neutral-beam injection.

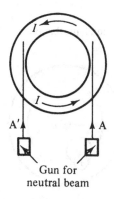

Gun for
neutral beam

antiparallel injection, by which beams are injected antiparallel to $I$, as designated by A′ in Fig. 2.16. Due to the narrowness of the open space around recently built large Tokomak machines, the neutral beams are in many cases injected obliquely or normally to the Tokomak. To extract neutral beams, a gaseous plasma is first formed by electric discharge in a box. A cathode of a net shape is set for the plasma, connected to the box with a biased voltage in the range $-100$ keV to $-1$ MeV. Only the positive ions are accelerated by this negative potential. The ions which pass through the cathode are charge-neutralised by the electrons of the gas near the cathode. Thus a neutral beam with an energy of 100 keV to 1 MeV is produced.

### 2.2.4. Wave heating

To initiate fusion reactions in the plasma, the ion temperature must be increased. In the method called wave heating, the ion temperature is increased by the electromagnetic waves which propagate in the plasma.

Ions in the Tokomak are constrained by strong magnetic fields so that they execute circular motions with the cyclotron frequency given by eq. (2.2). A coil is wound around the circular wall of the torus, with an alternating electric current of cyclotron frequency $\omega_{ci}$ flowing in the coil. The magnetic flux across the cross-section of the plasma column changes, inducing an alternating electric field in the azimuthal direction. The ion cyclotron motions resonate with this alternating electric field, and the kinetic energy of the ions with respect to the circular motions increases greatly (Fig. 2.17).

For the Tokomak, in addition to this ion cyclotron resonance, a lower hybrid wave is used to heat the plasma effectively (Fig. 2.18). Imagine that a magnetic field in the $z$-direction is applied to the uniform plasma with number density $n$. An electrostatic wave is

**Fig. 2.17.** Heating of ions in the plasma by ion cyclotron resonance.

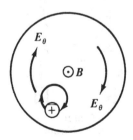

induced in the plasma with wave vector **k** in the $x$-direction and angular velocity $\omega$. The electric potential of this electrostatic wave is denoted by $\phi'$. The ion and electron velocities induced by this wave are denoted by $\mathbf{v}'_i$ and $\mathbf{v}'_e$, and the changes in the ion and electron number density $n'_i$ and $n'_e$, respectively. The equation of continuity and the equation of motion for the ion fluctuation are

$$\frac{\partial n'_i}{\partial t} + \nabla \cdot (n + n'_i) v'_i = 0, \qquad (2.69)$$

$$m_i \left\{ \frac{\partial v'_i}{\partial t} + (v'_i \cdot \nabla) v'_i \right\} = -e\nabla\phi' + ev'_i \times \mathbf{B}, \qquad (2.70)$$

respectively. In eq. (2.70), the ion pressure and the frictional force between the ion and the electron are neglected. Both equations may be linearised with respect to the fluctuating quantities, and reduced to

$$\frac{\partial n'_i}{\partial t} + n\nabla \cdot v'_i = 0, \qquad (2.71)$$

$$m_i \frac{\partial v'_i}{\partial t} = -e\nabla\phi' + ev'_i \times \mathbf{B}. \qquad (2.72)$$

The perturbed quantities are assumed to have a form of $e^{i(\omega t + kx)}$ with respect to time $t$ and space coordinate $x$, and the operations $\partial/\partial t$ and $\partial/\partial x$ are expressed by $i\omega$ and $ik$, respectively. We then have

$$n'_i = \frac{nkv'_{ix}}{\omega}, \qquad (2.73)$$

$$-i\omega m_i v'_{ix} = -iek\phi' + ev'_{iy}B, \qquad (2.74)$$

$$-i\omega m_i v'_{iy} = -ev'_{ix}B. \qquad (2.75)$$

**Fig. 2.18.** Propagation of the lower hybrid wave.

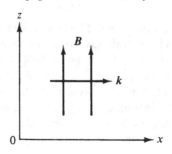

Here eq. (2.74) is the $x$-component equation and eq. (2.75) is the $y$-component equation of eq. (2.70). Equations (2.74) and (2.75) combine to give

$$v'_{ix} = \frac{ek}{m_i \omega} \phi' \left(1 - \frac{\omega_{ci}^2}{\omega^2}\right)^{-1}, \tag{2.76}$$

where $\omega_{ci} = eB/m_i$ is the ion cyclotron frequency. The same kind of equations as eqs (2.73) and (2.76) are derived for electrons as follows,

$$n'_e = \frac{nkv'_{ex}}{\omega}, \tag{2.77}$$

$$v'_{ex} = -\frac{ek}{m_e \omega} \phi' \left(1 - \frac{\omega_{ce}^2}{\omega^2}\right)^{-1}. \tag{2.78}$$

The Laplace equation for the potential $\phi'$ of the electrostatic wave is

$$\varepsilon \Delta \phi' = e(n'_i - n'_e). \tag{2.79}$$

In eq. (2.79), the dielectric constant $\varepsilon$ in the vacuum is so small that the right-hand side of eq. (2.79) is almost zero for a finite $\phi'$. The charge neutrality

$$n'_i = n'_e, \tag{2.80}$$

holds for many kinds of plasma. With the help of eq. (2.80), eq. (2.73) together with eq. (2.77) lead to

$$v'_{ix} = v'_{ex}.$$

Accordingly, the right-hand sides of eqs (2.76) and (2.78) are equal to each other and we can derive

$$m_i \left(1 - \frac{\omega_{ci}^2}{\omega^2}\right) = -m_e \left(1 - \frac{\omega_{ce}^2}{\omega^2}\right),$$

from which we get

$$\omega = (\omega_{ci} \omega_{ce})^{\frac{1}{2}}. \tag{2.81}$$

The frequency $\omega$ given by eq. (2.81) is called the lower hybrid frequency. The lower hybrid wave is an electrostatic wave which propagates perpendicular to the magnetic field. When a lower hybrid wave is launched to the plasma from outside the torus to disturb the plasma, ions in the plasma are resonantly fluctuated and increase the thermal energy. Heating by using the lower hybrid wave is one of the most promising methods of the secondary heating of plasma, as well as the means of sustaining the current in the Tokomak plasma.

When the vector of the wave number **k** is chosen to be parallel to the sub-axis of the torus, the lower hybrid wave can propagate along the sub-axis, since the poloidal magnetic field is perpendicular to **k**. The lower hybrid wave is a longitudinal wave (the electric fields of the wave are parallel to **k**). If the intensity of the wave is high enough, electrons in the plasma are captured by the tops of the electric potential of the wave. As a result, the captured electrons form a toroidal current. Thus the lower hybrid wave can be used to sustain the toroidal current. At present, the lower hybrid wave is considered the most promising method of sustaining the toroidal current in the Tokomak plasma, but the method has not yet been put to any practical use because of its low efficiency (Fig. 2.19).

The secondary-heating method, whether neutral beam heating or wave heating, must increase the plasma temperature about one order of magnitude. The energy of beams or waves brought into the plasma must be ten times the initial plasma energy. When such a large amount of energy is brought into the plasma, the initial state of the plasma confined by the magnetic field is disturbed significantly. Since there is a possibility that the confinement condition for the plasma will be changed by the input of the secondary heating energy, we must take into account the confinement of the plasma by changing its parameters.

### 2.2.5. Safety factors and the rotational transfer angle

In the stability condition (eq. (2.47)) for the spiral-type instability, $L$ is replaced by $2\pi R$ for the torus plasma. Therefore eq. (2.47) may be rewritten as

$$1 < \frac{aB_z}{RB_\theta(a)}. \tag{2.82}$$

Equation (2.82) describes the stability condition for $r = a$. However, the magnetic field must satisfy the same condition at an arbitrary

**Fig. 2.19.** Heating of the plasma by a lower hybrid wave.

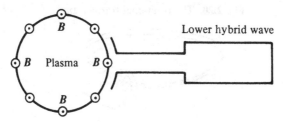

radius in the plasma column. The inequality

$$1 < \frac{rB_z}{RB_\theta(r)} = q_s(r), \qquad (2.83)$$

is the general stability condition for the spiral-type instability. In eq. (2.83), $q_s$ is called the safety factor. The plasma becomes more stable as $q_s$ increases, not only for the spiral-type instability but also for the sausage-type and the kink-type instabilities. From the experimental results, $2 < q_s$ is required for stability of the plasma.

When a space point moves one turn along a magnetic field line on the circular wall of the torus plasma around the principal axis, if $q_s$ is less than unity the space point moves less than one turn around the sub-axis. For such a case, the angle $\zeta$ given by

$$\zeta = \frac{2\pi}{m} \qquad (2.84)$$

specifies the angle at which the magnetic field line turns around the sub-axis, when the field line turns once around the principal axis. This angle $\zeta$ is called the rotational transfer angle. The number $m$ in eq. (2.84) is not limited to integers. When $m$ is an irrational number, a magnetic field line never comes back to the starting point regardless of the number of turns around the principal axis. Thus an endless field line describes a torus surface. This surface is called the magnetic surface (Fig. 2.20).

If we observe the magnetic field lines in the plasma column, the rotational transfer angle $\zeta$ is a function of radius $r$. In the Tokomak, the toroidal magnetic field $\mathbf{B}_z$ is almost constant regardless of the minor radius, while the poloidal field $\mathbf{B}_\theta$ increases with the minor radius $r$. Accordingly, inside the plasma column where $r$ is less than the minor radius $a$, $m$ in eq. (2.84), which defines the rotational transfer angle $\zeta$, becomes a rational number on surfaces with several minor radii. A magnetic surface with an irrational number $m$ is called an irrational surface; a surface with a rational number $m$ is called

**Fig. 2.20.** The rotational transfer angle.

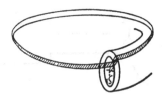

a rational surface. On the irrational surface, a perturbation does not grow easily and the plasma is generally stable there. But a perturbation does have a tendency to grow on the rational surface, so the plasma sometimes becomes unstable there.

The cylindrical coordinate system $(r, \theta, z)$ is used here for the plasma, as is shown in Fig. 2.21. The equation which describes a magnetic field line is given by

$$\frac{r \, d\theta}{dz} = \frac{B_\theta}{B_z}.$$

Integrating this equation around the major axis, we have

$$\zeta = \int d\theta = \int \frac{B_\theta}{rB_z} \, dz = \frac{B_\theta 2\pi R}{rB_z} = \frac{2\pi}{q_s}, \qquad (2.85)$$

which shows that $\zeta$ is inversely proportional to $q_s$. Since $B_\theta$ is a function of $r$, $q_s$ is also a function of $r$. The rotational transfer angle $\zeta$ is a function of $r$, too. This fact expresses in other words that the magnetic field has a shear. A shear in the magnetic fields plays a role in stabilising the confined plasma.

Magnetic field lines along the torus plasma are curved. Therefore charged particles which move along the magnetic field lines have drift motions perpendicular to the field lines and to the curvatures, as is described in section 2.1.1. The direction of the drift motion depends on the charge. In Fig. 2.22, the magnetic field is denoted by $B$ and the curvature by $R$. The positive charge thus drifts downward and the negative charge drifts upward. When the rotational transfer angle is zero, the positive charges are concentrated in the lower part of the plasma column, the negative charges in the upper part. As a result, an electric field is induced from the lower to the upper part. This electric field $\mathbf{E}$ is perpendicular to the magnetic field $\mathbf{B}$. Thus in the plasma a strong drift motion appears due to these electric and magnetic fields. The plasma will soon disperse through contact with the wall caused by this drift motion.

**Fig. 2.21.** Cylindrical coordinates for the plasma column.

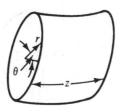

A charged particle undergoes drift motion due to centrifugal force when it moves along a magnetic field line with a curvature. The rotational transfer angle cancels this drift motion in the Tokomak. Electrons are concentrated first in the upper part of the plasma column, but are gradually shifted to the left part, to the lower part and to the right part as they move along the magnetic field lines. Thus the rotational transfer angle makes drift motions isotropic and prevents the plasma from charge separation; it thus contributes to confining the plasma in a stable way.

### 2.2.6. Diffusion of plasma

In order to continue nuclear fusion reactions it is necessary that plasma is confined over an interval of time, as indicated by the Lawson criterion in section 1.3.4. In magnetic confinement fusion, the plasma is confined by the Maxwell stress of the magnetic field. However, the plasma diffuses across the magnetic field.

Particles diffuse to make the gas density uniform if there is non-uniformity in the density of a gas. Since the mass flux density $\mathbf{I}$ is proportional to the gradient of the density $\nabla n$, we have

$$\mathbf{I} = -D\nabla n. \tag{2.86}$$

Here $D$ is the coefficient of diffusion. If the velocity of the diffusion is denoted by $v_{\text{dif}}$, the mass flux density $\mathbf{I}$ is given by

$$\mathbf{I} = nv_{\text{dif}}. \tag{2.87}$$

The equations of motion of the electron and the ion are

$$\rho_i \frac{\partial v_i}{\partial t} = -\nabla p_i + e(\mathbf{I}_i \times \mathbf{B}) + en_i \mathbf{E} - \frac{m_e}{\tau}(\mathbf{I}_i - \mathbf{I}_e), \tag{2.88}$$

$$\rho_e \frac{\partial v_e}{\partial t} = -\nabla p_e - e(\mathbf{I}_e \times \mathbf{B}) - en_e \mathbf{E} + \frac{m_e}{\tau}(\mathbf{I}_i - \mathbf{I}_e). \tag{2.89}$$

**Fig. 2.22.** Drift motion due to centrifugal force.

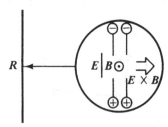

Here $\rho$ is the density, $p$ is the pressure and the suffixes i and e refer to the ion and the electron, respectively. The mean free flight time $\tau$ of an electron with respect to collisions with ions is given by eq. (2.65). The fourth terms in the right-hand sides of eqs (2.88) and (2.89) provide for the friction due to the relative velocity between the electron and the ion. The plasma is considered to be charge-neutral ($n_i = n_e = n$) and to be in a steady state. The left-hand sides of eqs (2.88) and (2.89) are then zero. The cylindrical coordinate system ($r, \theta, z$) is used. The magnetic field **B** is in the $z$-direction. The pressure $p$ is expressed by $p = nkT$, where the temperature $T$ is uniform and the number density $n$ is only a function of $r$. The $r$-components of eqs (2.88) and (2.89) give

$$-k(T_e + T_i)\frac{dn}{dr} + eB(I_{i\theta} - I_{e\theta}) = 0. \tag{2.90}$$

Equation (2.90) together with the $\theta$-components of eqs (2.88) and (2.89) lead to

$$I_{ir} = I_{er} = -\frac{m_e k}{e^2 B^2 \tau}(T_e + T_i)\frac{dn}{dr} + n\frac{E_\theta}{B}. \tag{2.91}$$

If we neglect the second term on the right-hand side of eq. (2.91), which is the drift term due to the electric field, there remains the term for the coefficient of diffusion $D_c$, given by

$$D_c = \frac{m_e k}{e^2 B^2 \tau}(T_e + T_i). \tag{2.92}$$

The coefficient of diffusion given by eq. (2.92) is called the coefficient of classical diffusion, and is proportional to $B^{-2}$. According to eq. (2.92), diffusion in the radial direction is reduced very much by increasing the intensity of the magnetic field.

The Larmor radius of an electron is given by

$$\rho_{re} = \frac{m_e v_\perp}{eB},$$

according to eq. (2.3). If the velocity $v_\perp$ is represented by the thermal velocity $(kT_e m_e)^{\frac{1}{2}}$, the Larmor radius is given by

$$\rho_{re} = \frac{(km_e T_e)^{\frac{1}{2}}}{eB}, \tag{2.93}$$

and combining with eq. (2.92),

$$D_c = \frac{\rho_{re}^2}{\tau}\left(1 + \frac{T_i}{T_e}\right), \tag{2.94}$$

is derived. The coefficient of classical diffusion is based on collisions of charged particles. A single charged particle rotating along the magnetic field with the Larmor radius cannot cross the magnetic field, but there are collisions between charged particles which rotate around the magnetic field lines. After a collision, the velocities of the two particles change, so a particle rotates around a different magnetic field line after a collision. The mean distance along which a guiding centre of the charged particle moves after a collision is of order of the Larmor radius. The coefficient of diffusion is given by the square of the mean distance travelled divided by the mean free flight time.

The change in the number density of the plasma is governed by the equation of diffusion

$$\frac{\partial n}{\partial t} = D_c \frac{\partial^2 n}{\partial r^2}, \qquad (2.95)$$

which is approximated by

$$\frac{n}{\tau_c} = D_c \frac{n}{a^2},$$

where $a$ is the minor radius of the Tokomak plasma. The confinement time $\tau_c$ of the plasma due to the classical diffusion is given by

$$\tau_c = \frac{a^2}{D_c} = \frac{a^2}{\rho_{re}^2} \tau, \qquad (2.96)$$

where eq. (2.94) is used.

Equation (2.96) shows that the confinement time is proportional to the square of the minor radius $a$ of the Tokomak plasma. The relation defining the parameters of the fusion plasma, depending on the scale of the facility, is called the scaling law. Equation (2.96) indicates that the experimental facility, in which fusion reactions of the plasma confined by magnetic fields are investigated, performs in the desired way when that facility is large. It must be said that the present Tokomaks which realise the critical condition are already of large scale, and a great amount of money was required to construct them. It is reasonable to suppose that a facility which extracts practical fusion energy will be larger still, and the money needed to construct it enormous. Whether such a large facility will supply electric power at a reasonable price is doubtful; it may be necessary to develop new ways of achieving nuclear fusion that involve facilities obeying a different scaling law.

Experimental results so far obtained show that the coefficient of diffusion $D_B$ is given semi-empirically by

$$D_B = \frac{1}{16} \frac{kT_e}{eB}, \tag{2.97}$$

which is proportional to $B^{-1}$, not $B^{-2}$. The coefficient of diffusion given by eq. (2.97) is called the coefficient of the Bohm diffusion. It has an astonishingly large value in comparison with that given by eq. (2.92).

The confinement time $\tau_B$ due to the Bohm diffusion is given by

$$\tau_B = \frac{a^2}{D_B},$$

and is called the Bohm time. The Bohm time is not dependent on the density $n$, as eq. (2.97) indicates. The coefficient of the Bohm diffusion given by eq. (2.97) increases with temperature $T$ contrary to the fact that the coefficient of the classical diffusion given by eq. (2.94) decreases with $T$ (through $\tau$). For the plasma with a temperature of 100 eV in the magnetic field of the intensity of 1 T, for instance, $D_B = 6 \text{ m}^2/\text{s}$ while $D_c = 5.5 \times 10^{-4} \text{ m}^2/\text{s}$. The fact that the Bohm diffusion is proportional to $B^{-1}$ suggests that the second term on the right-hand side of eq. (2.91) plays a more important role than the first term. In other words, we must take into account

$$I_{er} = \frac{nE_\theta}{B}. \tag{2.98}$$

If we consider that thermal motions separate the charge in the plasma and induce the electric potential, its maximum value may be estimated through

$$e\phi_{max} = kT_e.$$

The electric field $E_{max}$ of this potential is expressed approximately by using the minor radius $a$ of the plasma,

$$E_{max} = \frac{\phi_{max}}{a} = \frac{kT_e}{ea}.$$

Therefore, eq. (2.98) becomes

$$I_{er} = \frac{nkT_e}{eaB} = \frac{kT_e}{eB} |\nabla n| \approx D_B |\nabla n|, \tag{2.99}$$

which shows that $D_B$ is proportional to $T_e$ and $B^{-1}$.

In recent Tokomak experiments, confinement times of the order of $100\tau_B$ can be obtained by carefully reducing the induced fluctuating electric fields. If we succeed in reducing these fields in the plasma, the coefficient of diffusion is reduced from the large value given by the Bohm diffusion to the small one given by the classical diffusion. But in practice the coefficient of the classical diffusion must be modified a little.

The motion of a charge particle is constrained by the magnetic field. If the particle path follows the magnetic field line exactly, the particle path is on the magnetic surface. In reality, the particle executes the drift motion described in section 2.1.1. The curvature vector $\mathbf{K}$ of the magnetic field line is given by

$$\mathbf{K} = \frac{1}{B}(\mathbf{B} \cdot \nabla)\frac{\mathbf{B}}{B}.$$

Equations (2.9) and (2.10) give

$$\mathbf{v_D} = \frac{m(2v_\parallel^2 + v_\perp^2)}{2qB^3}(\mathbf{B} \times \nabla B). \tag{2.100}$$

An example of a path line of a particle in the $r$–$\theta$ plane is given by the dotted line in Fig. 2.23. The maximum distance $\delta$ of the deviation of the path line from the line of the magnetic surface is expressed by

$$\delta = \frac{2\pi}{\zeta}\rho_{\text{re}}. \tag{2.101}$$

Here $\rho_{\text{re}}$ is the electron Larmor radius and $\zeta$ is the rotational transfer angle. A charged particle changes its velocity in a collision, and as a result the path line shifts a distance of the order of $\delta$. Thus the

**Fig. 2.23.** The magnetic surface and drift surface of a charged particle in torus plasma.

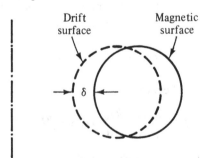

coefficient of diffusion due to this shift of path line is given by

$$D' = \frac{\delta^2}{\tau} = \frac{4\pi^2}{\zeta^2} D_c.$$

(2.102)

The coefficient of the classical diffusion given by eq. (2.92) added to $D'$ given by eq. (2.102) is

$$D = D_c\left(1 + \frac{4\pi^2}{\zeta^2}\right).$$

(2.103)

According to the more exact theory, the coefficient of diffusion $D_{p-s}$ due to particle collisions is derived to become

$$D_{p-s} = D_0\left(1 + \frac{8\pi^2}{\zeta^2}\frac{c_{e\perp}}{c_{e\parallel}}\right).$$

(2.104)

Here $c_e$ is the electrical conductivity, whose parallel component $c_{e\parallel}$ is given by eq. (2.67). The component of the electrical conductivity $c_{e\perp}$ perpendicular to the magnetic field depends on the intensity of the magnetic field and the collision frequency, and is roughly given by $c_{e\perp} = 3.3c_e$. The factor $(1 + 8\eta^2 c_{e\perp}/\zeta^2 c_{e\parallel})$ appearing in eq. (2.104) is called the Pfirsch–Schlütter factor, and shows the amplification factor of the coefficient of classical diffusion. The coefficient $D_{p-s}$ given by eq. (2.104) is called the coefficient of neoclassical diffusion.

Since the Tokomak plasma has a torus form, the intensity of the magnetic field is non-uniform. The intensity of the magnetic field at A in Fig. 2.24 inside the torus is strong, while that at A' outside the torus is weak. Because a charged particle moves almost exactly along a magnetic field line, the particle moves toward the direction along which the magnetic field intensity increases; so the particle executes a spiral motion from outside to inside the torus. As section 2.1.1 shows, the particle is subject to a decelerating force when it moves

**Fig. 2.24.** Magnetic field lines and banana path of a particle.

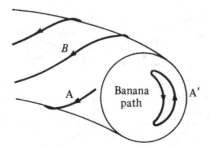

along a magnetic field line whose intensity is increasing. Hence charged particles with small initial velocities cannot pass the point of maximum intensity of the magnetic field, and so bounce on the way. The trajectory of the path line of such a bouncing particle on the $r-\theta$ plane is shown in Fig. 2.24. Since the shape of the trajectory resembles that of a banana, the trajectory is called the banana path. If there are particles which follow banana paths, the particle paths will change from banana paths to circular paths and vice versa, as a result of particle collisions. Thus a deviation of a particle path due to a collision becomes larger than $\rho_{re}$ or $\delta$, described above. Given this fact, the coefficient of diffusion $D$ is as shown in Fig. 2.25. The abscissa shows the collision frequency $\nu = 1/\tau$. In the region where the collision frequency is large, the coefficient $D$ coincides with the neoclassical one given by eq. (2.104). In the banana region, where the collision frequency is small and many particles have banana paths, the coefficient of diffusion $D_{ba}$ is given by

$$D_{ba} = \frac{q_s^2 \rho_{ri}^2}{\varepsilon_a^{\frac{3}{2}} \tau_b}. \tag{2.105}$$

Here $\varepsilon_a$ is the ratio of the minor radius of the Tokomak plasma to the major radius of it, $\varepsilon_a = a/R$ (where $\varepsilon_a$ is the aspect ratio of the torus plasma) and $q_s$ is the safety factor, $q_s = 2\pi/\zeta$, where $\zeta$ is the rotational transfer angle. In eq. (2.105), $\tau_b$ is the time interval during which a particle moves along a banana path. In the intermediate region, called the plateau region, $D_p$ is almost constant and is given by

$$D_p = \frac{q_s^2 \rho_{ri}^2}{\varepsilon_a^{\frac{1}{2}} \tau_b}. \tag{2.106}$$

**Fig. 2.25.** Coefficient of diffusion and collision frequency.

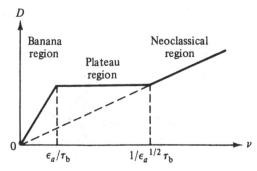

### 2.2.7. *Instabilities in the Tokomak plasma*

There are several instabilities that have an important influence on the confined Tokomak plasma, in addition to the sausage-type, kink-type, and the spiral-type instability described in sections 2.1.3–5. Instabilities are classified into two groups: macro-instability, which is analysed on the basis of hydrodynamic treatment; and micro-instability which is associated with the velocity distribution of the electrons or ions.

The flute-type instability must be considered first among the macro-instabilities. If the plasma has a convex surface toward the magnetic field, as in Fig. 2.26, centrifugal force acts on charged particles which move nearly along the magnetic field lines. Since the centrifugal force is perpendicular to the magnetic field, charged particles execute drift motion as described in section 2.1.1. Suppose that the perturbation appears on the boundary between the plasma and the magnetic field, as in Fig. 2.27. The direction of drift motions differs according to the charge of the particles, and charge separation occurs on the boundary due to these differences. The charge

**Fig. 2.26.** Plasma surrounded by a magnetic field.

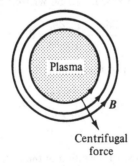

Centrifugal
force

**Fig. 2.27.** Flute-type instability.

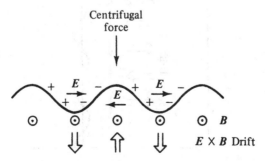

separation induces locally-fluctuating electric fields, which cause drift motions due to $E \times B$ and so enhance the perturbation on the boundary. Thus the perturbation is unstable. Perturbation will build up ditches along the magnetic field line; this instability is called flute-type instability because of its shape. It is necessary that the plasma does not have a convex form toward the magnetic field if flute-type instability in the plasma is to be prevented.

Next, let us investigate drift-wave instability. The magnetic field around the Tokomak plasma is nearly in the $z$-direction. The density of the plasma confined by the magnetic field decreases with $r$, if the cylindrical coordinate system $(r, \theta, z)$ is used. Here we suppose that the fluctuation $\delta n_i$ of the ion density occurs in the plasma along the $\theta$-direction. This fluctuation of the ion density induces the fluctuation of the electric potential $\phi$. The electric field $E_\theta$ in the $\theta$-direction caused by this fluctuation of $\phi$ is given by $E_\theta = -\partial\phi/r\,\partial\theta$. Charged particles have drift motions due to $E_\theta$ and $B_z$. This drift motion conveys the fluctuation $\delta n_i$ of the ion density along the $\theta$-direction. This propagating wave is called the drift wave. The fluctuation $\delta n_i$ of the ion density is expressed by $\delta n_i \propto \sin(kr\theta - \omega t)$, and the equation of continuity for the fluctuation $\delta n_i$ of the ion density for the drift motion due to $E \times B$ can be written as

$$\frac{\partial \delta n_i}{\partial t} + \frac{E_\theta}{B} \frac{\partial n_i}{\partial r} = 0. \tag{2.107}$$

The equation of motion in the $\theta$-direction is

$$n_i E_\theta + \frac{kT}{m_i} \frac{\partial \delta n_i}{r\,\partial\theta} = 0. \tag{2.108}$$

**Fig. 2.28.** Propagation of the drift wave.

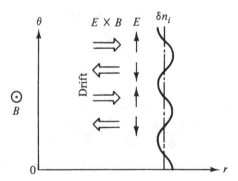

Equations (2.107) and (2.108) may be combined to give the dispersion relation

$$\omega = k \frac{k}{m_i} \frac{T}{eB} \left| \frac{\partial n_i}{n_i \, \partial r} \right|, \qquad (2.109)$$

where $\omega$ is called the drift frequency.

The drift wave frequently becomes unstable, due to several causes. For instance, consider the following cases. The ion drift motion sometimes begins after the electron drift motion, due to the large mass or the large Larmor radius of the ion in comparison with those of the electron (although the $\mathbf{E}_\theta \times \mathbf{B}_z$ drift motions are the same for the electron and the ion in the steady state). For such a case, the fluctuation $\delta\phi$ of the electric potential has a phase lag in the fluctuation $\delta n_i$ of the ion density. The drift motion bases on $\mathbf{E}_\theta \times \mathbf{B}_z$, where $\mathbf{E}_\theta$ has a phase lag and enhances the fluctuation $\delta n_i$ of the ion density, as in Fig. 2.29. The drift-wave instability exerts a large influence on the plasma confinement by increasing the transport coefficients, but it seems that the increase in the shear of the magnetic fields in the Tokomak plasma suppresses such drift-wave instability.

The temperature of the fusion plasma is very high. The collisions among charged particles are not so frequent at such a high temperature, so macroscopic analysis of the plasma using MHD equations is not valid for the fusion plasma. From the point of view of the kinetic theory, the Vlasov equations

$$\frac{\partial f_i}{\partial t} + \mathbf{v} \cdot \frac{\partial f_i}{\partial \mathbf{r}} + \frac{e_i}{m_i} (\mathbf{E} + \mathbf{v} \times \mathbf{B}) \cdot \frac{\partial f_i}{\partial \mathbf{v}} = 0, \qquad (2.110)$$

$$\frac{\partial f_e}{\partial t} + \mathbf{v} \cdot \frac{\partial f_e}{\partial \mathbf{r}} - \frac{e}{m_e} (\mathbf{E} + \mathbf{v} \times \mathbf{B}) \cdot \frac{\partial f_e}{\partial \mathbf{v}} = 0, \qquad (2.111)$$

**Fig. 2.29.** Growth of the drift wave.

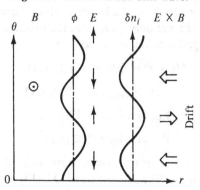

for the ion velocity distribution $f_i$ and the electron velocity distribution $f_e$, combined with the Maxwell equations

$$\text{rot } \mathbf{E} + \frac{\partial \mathbf{B}}{\partial t} = 0, \tag{2.112}$$

$$\frac{1}{\mu} \text{rot } \mathbf{B} - \frac{\varepsilon \, \partial \mathbf{E}}{\partial t} = \int (e_i \mathbf{v} f_i - e \mathbf{v} f_e) \, d\mathbf{v}, \tag{2.113}$$

for the electric field $\mathbf{E}$ and the magnetic field $\mathbf{B}$, are the more pertinent governing equations. It turns out that there are many micro-instabilities which are obscure under hydrodynamic analysis for macroscopic variables only, but which become clear if they are treated according to kinetic theory.

As an example of micro-instability, trapped-particle instability will be described here. Because of the toroidal form of the Tokomak plasma, the intensity of the magnetic field inside the torus is greater than that outside the torus, as explained in section 2.2.6. Therefore some charged particles (electrons or ions) bounce in the inner part of the torus and move along banana paths. These particles are called the trapped particles. The average frequency $\omega_b$ of the trapped particles which move along banana paths is smaller than the mean collision frequency if the temperature of the plasma is very high. Then $\omega_b$ is the characteristic frequency of the plasma in the banana region. If there is a perturbation whose frequency coincides with $\omega_b$, this perturbation in the plasma becomes unstable, resonating with the trapped particles and increasing transport coefficients. The effects of trapped-particle instability in the banana region on the coefficients of thermal conductivity and of the diffusion are currently being studied.

Using the cylindrical coordinate $\mathbf{r}(r, \theta, z)$, let us now investigate the stability of the Tokomak plasma. In the magnetically-confined steady plasma, the variables which specify plasma conditions do not depend on $t$, $\theta$ and $z$ but only on $r$. Here the variables are separated into steady and perturbed terms as follows,

$$f_i(t, \mathbf{v}, \mathbf{r}) = f_{i0}(\mathbf{v}, r) + f'_i(t, \mathbf{v}, \mathbf{r}), \tag{2.114}$$

$$f_e(t, \mathbf{v}, \mathbf{r}) = f_{e0}(\mathbf{v}, r) + f'_e(t, \mathbf{v}, \mathbf{r}), \tag{2.115}$$

$$\mathbf{E}(t, \mathbf{r}) = \mathbf{E}_0(r) + \mathbf{E}'(t, \mathbf{r}), \tag{2.116}$$

$$\mathbf{B}(t, \mathbf{r}) = \mathbf{B}_0(r) + \mathbf{B}'(t, \mathbf{r}). \tag{2.117}$$

Here the suffix 0 refers to the steady state while the prime refers to the perturbations. Equations (2.110)–(2.113) are replaced by eqs (2.114)–(2.117), and linearised with respect to the perturbed quantities. Moreover, the perturbed quantities are represented by Fourier components with respect to $\theta$, $z$ and $t$ as follows:

$$f_i'(t, \mathbf{v}, \mathbf{r}) = f_i^*(\mathbf{v}, r) \exp i(m\theta + nz - \omega t), \tag{2.118}$$

$$f_e'(t, \mathbf{v}, \mathbf{r}) = f_e^*(\mathbf{v}, r) \exp i(m\theta + nz - \omega t), \tag{2.119}$$

$$\mathbf{E}'(t, \mathbf{r}) = \mathbf{E}^*(r) \exp i(m\theta + nz - \omega t), \tag{2.120}$$

$$\mathbf{B}'(t, \mathbf{r}) = \mathbf{B}^*(r) \exp i(m\theta + nz - \omega t). \tag{2.121}$$

The system of the linearised equations then constitutes an eigenvalue problem for the perturbations. If we give small amplitudes $f_i^*(\mathbf{v}, 0)$, $f_e^*(\mathbf{v}, 0)$, $\mathbf{E}^*(0)$ and $\mathbf{B}^*(0)$ for the perturbations at the sub-axis of the Tokomak $r = 0$, and set the amplitudes of the perturbations at zero on the surface of the minor radius of the Tokomak plasma as follows, $f_i^*(\mathbf{v}, a) = 0$, $f_e^*(\mathbf{v}, a) = 0$, $\mathbf{E}^*(a) = 0$ and $\mathbf{B}^*(a) = 0$, then we can obtain $f_i^*(\mathbf{v}, \mathbf{r})$, $f_e^*(\mathbf{v}, \mathbf{r})$, $\mathbf{E}^*(\mathbf{r})$ and $\mathbf{B}^*(\mathbf{r})$ as the eigenfunctions, and the dispersion relation

$$\omega = \omega(m, n) \tag{2.122}$$

is derived as an eigenvalue. The complex value of $\omega$ is separated into two as

$$\omega(m, n) = \omega_r(m, n) + i\gamma(m, n), \tag{2.123}$$

where $\omega_r$ stands for the frequency (angular velocity) of the perturbation and $\gamma$ stands for the growth rate. When $\gamma > 0$, the steady state of the Tokomak is unstable for the perturbation of the mode $m/n$. When $\gamma > 0$ for the mode $m/n = 1/0$, the sausage-type instability occurs, and when $\gamma > 0$ for the mode $m/n = 1/1$, the kink-type instability occurs. The stability of the Tokomak plasma depends essentially on the profile of the toroidal current, i.e. on $f_{e0}^*(\mathbf{v}, r)$, and the plasma is sometimes unstable for the modes $m/n = 2/1$ and $m/n = 3/2$. Since the electron current flows in the plasma, the Tokomak plasma is likely to be unstable in comparison with the plasma lacking the current, because the existence of the rational surfaces in the plasma initiates the perturbations. If the nonlinear coupling of the two modes $m/n = 2/1$ and $m/n = 3/2$ are linearly unstable, the result shows that the perturbations grow rapidly to violate the steady state of the Tokomak plasma. This phenomenon is called disruptive instability. When disruptive instability occurs in the Tokomak, the toroidal current is shut down abruptly, and the Tokomak may be damaged. Every mode $m/n$ of perturbations should be stable, or should be stabilised by the feedback system in the plasma which is confined steadily.

### 2.2.8. The present state of Tokomak research

The ordinate of Fig. 2.30 gives the fusion parameter $n\tau$ and the abscissa the plasma temperature $T$. The plasma conditions achieved by Tokomaks already constructed are indicated by ● and those of the critical Tokomaks which are in operation or under construction by ○. The characteristics of the four critical Tokomaks, the JT-60 at Naka of the Japanese Atomic Energy Research Institute, the TFTR at Princeton University in the USA, the JET at the Culham Laboratory of EURATOM, and the T-15 in the Kurchatov Institute in the USSR, are tabulated in Table 2.2. The aim of these four Tokomaks is to achieve the Lawson criterion at the lowest level.

## 2.3. Magnetic-confinement fusion using devices other than the Tokomak

### 2.3.1. Reversed-field pinch

Electrons form the major part of the toroidal current in the Tokomak plasma. Roughly speaking, electrons move along the magnetic field lines. The magnetic field in the Tokomak has a poloidal component, because the plasma is confined by the poloidal magnetic field. Thus the electron current in the Tokomak plasma also has a poloidal component (Fig. 2.31). The poloidal current of electrons induces a

**Fig. 2.30.** Stages in the development of the Tokomak.

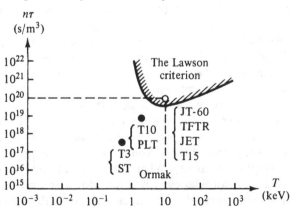

Table 2.2 *Critical Tokomaks*

| Facility | JT-60 | JET | TFTR | | T-15 |
|---|---|---|---|---|---|
| Location | Naka Japan JAERI | Calham UK EURATOM | Princeton USA | | Kurchatov USSR |
| Major radius (m) | 3 | 2.96 | 2.48 | | 5 |
| Minor radius (m) | 1 | 1.25 | 0.85 | | 2 |
| Vertical radius of the plasma (m) | 1 | 2.10 | 0.85 | | 2 |
| Aspect ratio | 3 | 2.37 | 2.9 | | 2.5 |
| Intensity of the magnetic field (T) | 5 | 3.4 | 5.2 | | 3.5 |
| Plasma current (MA) | 3.3 | 4.8 | 2.5 | 1.0 | 6 |
| Safety factor at the surface | 3.5 | 6 | 3.0 | 7.5 | 2.3 |
| Average ion temperature (keV) | 5–10 | 5 | 6.0 | 6.0 | 7–10 |
| Energy confinement time (s) | 0.2–1 | 1 | 0.2 | | 2 |
| Volume of the vacuum vessel ($m^3$) | 100 | 190 | 64 | | 400 |
| Power of the neutral beam injection (MW) | 10–20 | 3–25 | 12 | 40 | 60 |
| RF heating power (MW) | 10 | 3–20 | | | 60 |
| RF frequency (MHz) | 1000 | 800 | | | 1000 |

**Fig. 2.31.** Reversed-field pinch.

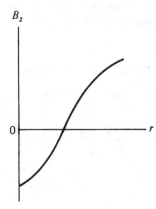

toroidal magnetic field which has the opposite direction to the field induced by the external coils. Therefore the toroidal magnetic field near the sub-axis of the Tokomak is reversed if the toroidal current increases over the Kruskal–Shafranov limit, and the poloidal current of electrons increases. A machine to confine magnetically the torus plasma which has a reversed toroidal magnetic field is called the reversed-field pinch. The $\beta$-value of the Tokomak plasma is small since the strong toroidal magnetic field is applied to the plasma to stabilise it, thus the Tokomak is a machine of low economy. By contrast, plasma in the reversed-field pinch can have a large $\beta$-value. Investigations at the plasma in the reversed-field pinch focus on its configuration at equilibrium and its stability.

### 2.3.2. *The stellarator and the heliotron*

The poloidal magnetic field in the Tokomak is applied by the toroidal current, which is induced in the torus plasma as the secondary coil of the transformer when the current in the primary coil increases. It is impossible for the secondary current in the transformer (toroidal current in the plasma) to flow steadily. The poloidal magnetic field can be applied by a current flowing in an external coil of a spiral shape surrounding the plasma. Alternating electric currents flowing along the six spiral coils (denoted by $l = 3$; the number of coils is generally even) as shown in Fig. 2.32 form the magnetic fields of a triangle shape near the sub-axis of the torus. There are strong magnetic shears on the magnetic surfaces inside the plasma. The magnetic fields induced by this method are known as stellarator fields. The currents along the spiral coils must not alternate; if the halves of the coils in Fig. 2.32 are removed alternately, similar magnetic fields can be formed. Such magnetic fields are called the heliotron fields. Currents do not flow in the stellarator or the heliotron plasma, so for them steady operation will be possible if

**Fig. 2.32.** The Stellarator.

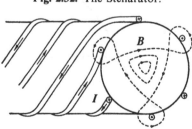

the coils are super-conducting. The magnetic fields are non-symmetric along the azimuthal direction of the minor radius.

### 2.3.3. The Miller field

Charged particles which move along magnetic fields and rotate around them are accelerated toward the directions along which the magnetic fields increase, as described in section 2.1.6. Charged particles, gyrating around the magnetic fields induced by the two coils along which currents flow in the same direction (Fig. 2.33), can be confined in the central part between the two regions A and A', where the coils are located and the intensities of the magnetic fields are strong. The ratio of the particle velocity component $v_\perp$ perpendicular to the magnetic field to the total velocity $v$ is denoted by $\sin \theta$. The angle $\theta$ at the point where the magnetic field is $B_0$ is denoted by $\theta_0$. Since the magnetic moment $mv_\perp^2/2B$ of the charged particle is constant (section 2.1.6), $v_\perp^2$ is proportional to $B$. The particle speed $v$ is constant because the magnetic field does no work on the charged particle. Thus we have

$$\sin^2 \theta = \frac{B}{B_0} \sin^2 \theta_0. \qquad (2.124)$$

At the point where $B$ is large and $B/B_0$ is equal to $1/\sin^2 \theta_0$, $v_\perp$ becomes equal to $v$ and the particle velocity component $v_\parallel$ parallel to the magnetic field is zero. The charged particle bounces at this point and returns to the direction along which the magnetic field decreases. The maximum value of the magnetic field is denoted by $B_m$. If the inequality

$$v_\perp^2/v^2 = \sin^2 \theta_0 > B_0/B_m = \sin^2 \theta_{min} \qquad (2.125)$$

is satisfied, the charged particle is confined in the central region

Fig. 2.33. The Miller field.

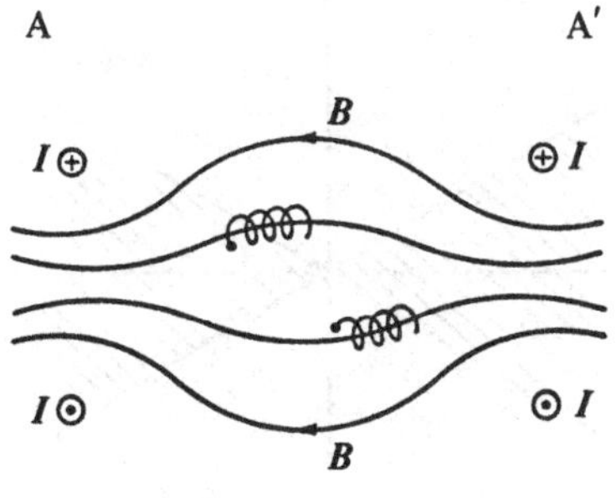

between the two coils, bouncing on the barriers of the strong magnetic fields. A magnetic field with neck parts at which charged particles bounce is called a Miller field. Such Miller fields can be used to confine charged particles. Charged particles whose velocities are located in the dotted regions in Fig. 2.34 are confined by the Miller field, since all charged particles whose velocities satisfy the inequality (2.125) are confined. Neutral beams with high velocity are injected into the central part of the Miller field perpendicular to the magnetic field; the particles obtain charges, depriving the ions in the plasma of the charges, and are confined in the Miller field with large $v_\perp$ and with small $v_\parallel$. If the density of ions with such high velocities increases in the Miller field, the plasma condition will exceed the Lawson criterion. Collisions among charged particles increase when the density of the charged particles increases in the Miller field. The velocity of a particle changes after a collision; this particle velocity after a collision happens to be located in the hatched regions in Fig. 2.34. Particles whose velocities are located in the hatched region are not confined by the Miller field. The hatched region of the velocity space of charged particles in the Miller field is called the loss cone. Charged particles, whose velocity is transferred into the loss cone, run away from the Miller field, passing through its neck, where the intensity of the magnetic field is at a maximum. The kinetic energies of the runaway particles can be converted into electrical energy through the direct energy conversion method of MHD.

### 2.3.4. The cusp field

The cusp field is a magnetic field which is induced by the two coils of the Miller field in the case where the current in one coil flows in

**Fig. 2.34.** Loss cone for charged particles in the Miller field.

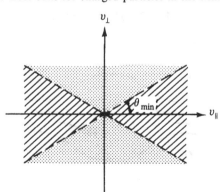

the opposite direction to that in the other coil, as in Fig. 2.35. At the cusp points P and Q, the charged particle with small $v_\parallel$ bounces because the intensity of the magnetic field increases when the particle approaches the cusp point from the central part. At the cusp ring S and T, by contrast, the magnetic field is open and charged particles can escape from the cusp field. The cusp field therefore seems to be ineffective for confinement of the plasma. From the point of view of the exchange-type instability, however, the cusp field is more effective in confining charged particles than the Miller field.

Let us consider two magnetic flux tubes $A_1$ and $A_2$ (Fig. 2.36) which neighbour each other in the plasma. Both tubes are small enough that the physical quantities in a cross-section can be assumed to be uniform. If the length, the cross-sectional area, the volume, and the magnetic flux density of the tube, respectively, are referred to $l$, $S$, $V$ and $B$, we have

$$V = \int S \, dl = \Phi \int \frac{dl}{B}, \tag{2.126}$$

where $\Phi$ is the magnetic flux in the tube and $\Phi = BS = $ const. The energy $E_m$ in the tube is given by

$$E_m = \int \frac{B^2}{2\mu} \, dV = \frac{\Phi^2}{2\mu} \int \frac{dl}{S}. \tag{2.127}$$

The two magnetic flux tubes $A_1$ and $A_2$ are chosen such that the magnetic flux $\Phi$ and $\int dl/S$ in the tubes are equal to each other. When the plasma in tube $A_1$ is exchanged for that in tube $A_2$, the

**Fig. 2.35.** Cusp field.

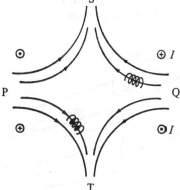

magnetic flux $\Phi$ and the energy $E_m$ of the magnetic field remain unchanged. Let us denote the pressure and the density in the tube $A_1$ before the exchange by $p_1$ and $\rho_1$, and the pressure and the density in the tube $A_2$ before the exchange by $p_2$ and $\rho_2$. The corresponding values after the exchange are denoted by $p'_1$, $\rho'_1$, $p'_2$ and $\rho'_2$, respectively. If the exchange is carried out adiabatically, we have

$$\frac{p'_1}{p_2} = \left(\frac{V_2}{V_1}\right)^\gamma, \qquad \frac{p'_2}{p_1} = \left(\frac{V_1}{V_2}\right)^\gamma.$$

In the above equations, $V_1$ and $V_2$ are the volumes of $A_1$ and $A_2$, and $\gamma$ is the ratio of the specific heat at a constant pressure to the specific heat at a constant volume; it is called the adiabatic constant. The internal energy of the gas in the unit volume is given by

$$e = \rho C_v T = \frac{C_v p}{R} = \frac{C_v p}{C_p - C_v} = \frac{p}{\gamma - 1}.$$

The change $\Delta E_1$ of the internal energy, when the plasma in tube $A_1$ is exchanged for that in tube $A_2$, is

$$\Delta E_1 = (p'_1 - p_1)\frac{V_1}{\gamma - 1} + (p'_2 - p_2)\frac{V_2}{\gamma - 1}$$

$$= \left\{ p_2\left(\frac{V_2}{V_1}\right)^\gamma - p_1 \right\}\frac{V_1}{\gamma - 1} + \left\{ p_1\left(\frac{V_1}{V_2}\right)^\gamma - p_2 \right\}\frac{V_2}{\gamma - 1},$$

which reduces to

$$\Delta E_1 = \Delta V\left(\Delta p + \frac{\gamma p_1}{V_1}\Delta V\right),$$

where $p_2 - p_1 = \Delta p_1$, $V_1 - V_1 = \Delta V$; the third and higher order terms of the small quantities are neglected. Since the exchange of the plasma

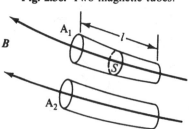

**Fig. 2.36.** Two magnetic tubes.

is carried out adiabatically and the energy of the magnetic field remains constant after the exchange, the plasma before the exchange is stable if $\Delta E_1 > 0$, and unstable if $\Delta E_1 < 0$. Now tube $A_1$ is in the vacuum outside the plasma and tube $A_2$ is inside the plasma. Then $p_1 = 0$, $p_2 \neq 0$ and hence $\Delta p > 0$. Thus we have

$$\Delta E_1 = \Delta V \Delta p.$$

The plasma before the exchange is stable if $\Delta V \propto \Delta \int dl/B > 0$ and the plasma is unstable if $\Delta V \propto \Delta \int dl/B < 0$.

If the magnetic field is convex toward the outside as in the Miller field, its intensity is weaker in the outer region, and the cross-sectional area of the outer magnetic flux tube is larger than that of the inner one for the same $\Phi$. Since the two tubes are supposed to have the same $E_m$, the outer tube must be longer than the inner one. Equation (2.126) shows that $B$ is smaller and $l$ is larger for the outer tube. Thus the volume of the outer tube is larger and $\Delta V$ is negative. The plasma in the Miller field turns out to be unstable. The same kind of consideration leads to the conclusion that the plasma in a cusp field which is concave toward the outside is stable with regard to exchange-type instability. Exchange-type instability is the alternative name of the flute-type instability described in section 2.2.7. For confinement of plasma to be effected using the cusp field, it is required to suppress the leaking of charged particles from the cusp ring parts S and T. Plugging the cusp ring by applying a RF alternating electric field is being examined as a way of suppressing such leaking.

### 2.3.5. The baseball field

The plasma is stable with regard to the exchange-type instability if the intensity of the magnetic field surrounding the plasma is stronger in the outer part regardless of direction. This kind of configuration of the magnetic field is called the minimum-$B$ field or the Ioffe field.

**Fig. 2.37.** Currents to induce the minimum-$B$ field.

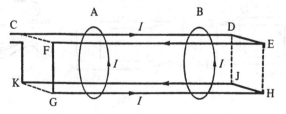

The minimum-*B* field can be formed when the electric currents flow in the two Miller coils A and B in the same direction, and flow in alternate directions in the four straight wires CD, EF, GH and JK, which are perpendicular to the Miller coils A and B (Fig. 2.37). The currents shown in Fig. 2.37 are redrawn schematically in Fig. 2.38(a), which are effectively equal to those in Fig. 2.28(b). If the current flows along the seam of the baseball, then the minimum-*B* field is formed to confine the plasma efficiently (see Fig. 2.38(c)).

### *2.3.6. The bumpy torus*

The Miller field, the cusp field and the baseball field may for simplicity be called the Miller field, generally speaking. The volume of the plasma confined by this Miller field can increase if the distance between the Miller coils increases. In a fusion reactor using plasma confined by a magnetic field, the wall loading is severe. Damage to the first wall by neutrons, etc., will be an important problem in the future; it will be explained in Chapter 4. The Miller field offers the possibility of reducing the wall loading. On the other hand, the Miller field has the disadvantage that particles in the loss cone leak out along the magnetic field lines. To eliminate this disadvantage, several Miller fields are connected to form a torus called the bumpy torus. Since the plasma in the bumpy torus is unstable macroscopically, efforts have been made to stabilise it by forming electron ring currents in the plasma in the central parts between the Miller coils.

### *2.3.7. The compact torus*

Although the iron yoke of the transformer has been removed from the principal axis in recent large Tokomaks, the vicinity of the principal axis of the Tokomak is crowded by the coils for the toroidal field and other materials. This jam near the principal axis causes

**Fig. 2.38.** Baseball coil.

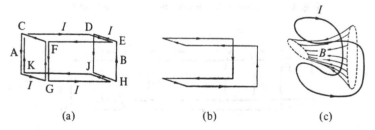

(a)                    (b)                    (c)

the structural complexity of the Tokomak. If the neighbourhood of the principal axis of the torus plasma is in a vacuum or is filled by diluted plasma with no other material present, the torus plasma is generally called the compact-torus plasma. There are several ways of constructing such a compact-torus plasma. One way is as follows. Plasma of a ring shape is launched by the plasma gun into the cusp field from the outside. When the plasma moves with the velocity $v_z$ in the $z$-direction in a magnetic field **B** with the radial component $B_r$, the Lorentz force

$$E_\theta = v_z B_r$$

acts on the ring plasma. Electrons in the plasma form the toroidal current $j_\theta$ due to the Lorentz force, and the current $j_\theta$ induces a magnetic field $B_\varphi$ around the compact torus plasma (Fig. 2.39). Once the compact torus plasma is formed, the magnetic field **B** which is applied externally is modified to guide the torus plasma to the position where the output fusion energy is to be extracted from the plasma. The objective of the devices for confining the plasma by magnetic fields described in the preceding paragraphs is that they operate steadily. On the other hand, the compact torus seems to operate quasi-steadily. After the fusion output energy is released from one compact torus plasma, the next plasma is formed successively. Thus the compact torus will operate with a small repetition rate.

### 2.3.8. The theta-pinch machine

The devices for magnetic confinement described in this and subsequent paragraphs are considered to be pulse-operated. The Lawson criterion will be satisfied by the condition that the plasma

**Fig. 2.39.** Compact torus.

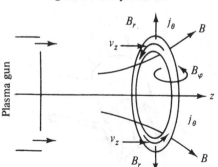

of number density of $n = 10^{21}$–$10^{23}$ m$^{-3}$ is magnetically confined over a time interval of $\tau = 1$–100 ms. The scaling laws which apply to these devices are expected to be significantly different from that of a Tokomak which aims to operate steadily.

A one-turn coil of the slab type surrounds a cylinder, which is filled by plasma (Fig. 2.40). When a giant current suddenly flow in the coil, a strong magnetic field in the axial direction is induced in the gap between the coil and the plasma. This magnetic field compresses the plasma to a high density at a high temperature. This type of device, which confines the plasma tentatively, is called the theta-pinch machine. A magnetic field whose direction is the same as or the opposite of that induced by the one-turn coil is usually applied to the plasma before the current flows in the coil. The theta-pinch machine has the disadvantage that the plasma runs away along the axis toward the two end sections, being compressed radially at the central part of the cylinder. If the plasma has a biased field whose direction is opposite to that of the coil, however, the magnetic field is reconnected with the field induced by the coil to form a ring line, as in Fig. 2.41. This ring line of the magnetic field suppresses the axial motion of the plasma.

Figure 2.42 shows schematically the cross-section of the cylinder. The strong magnetic field $\mathbf{B}_0$ which is induced by the current $I$ in the coil compresses the plasma. If the rise-time of the magnetic field is very short, the magnetic field cannot penetrate the plasma and a surface current $j_s$ is induced in the azimuthal direction on the plasma surface $C$. Since the plasma surface is suddenly accelerated radially, a shock wave appears in the plasma, propagating with a velocity $v'_s$ toward the centre O. The compressed and heated plasma flows with velocity $v'_2$ behind the shock wave. The plasma inside the shock

**Fig. 2.40.** Theta-pinch.

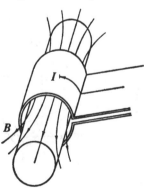

wave $S$ remains undisturbed. The plasma velocity is zero and the magnetic field $\mathbf{B}_0$ does not affect this region. It is easier for us to study the phenomenon in the slab configuration. In Fig. 2.43, the region ① is the initial state, which remains undisturbed. The density, pressure, temperature, magnetic field, and velocity there are denoted by $\rho_1$, $p_1$, $T_1$, $B_1$, and $v_1' = 0$, respectively. Here $B_1$ is the biased field. The shock wave $S$ propagates toward region ① with the velocity $v_s'$. Region ② shows the plasma which is compressed and heated by the shock wave. The density, pressure, temperature, magnetic field, and velocity there are denoted by $\rho_2$, $p_2$, $T_2$, $B_2$ and $v_2'$, respectively. The biased field $B_1$ in front of the shock wave is compressed to $B_2$ behind the shock wave. The boundary between the vacuum and the plasma is moving with velocity $v_2'$ and stands by C, on which the surface current $J_s$ flows. The vacuum region ③ is filled by the magnetic field $B_0$ which is induced by the coil. On both sides of the boundary C, the pressure (including the Maxwell stresses) is

**Fig. 2.41.** Reconnection of the magnetic field in the theta-pinch machine.

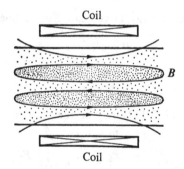

**Fig. 2.42.** A shock wave propagating in the plasma in the theta-pinch machine.

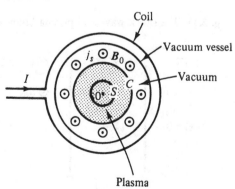

balanced. Thus we have

$$p_2 + \frac{B_2^2}{2\mu} = \frac{B_0^2}{2\mu}. \tag{2.128}$$

One of the Maxwell equations leads to

$$J_s = B_0 - B_2 \tag{2.129}$$

for $J_s$. Let us Galilei-transform the coordinate system to that in which the shock wave is at rest. In the new coordinate system, the velocity in the region ① is denoted by $v_1$, and that in region ② by $v_2$. Then we have $v_1 = v_s'$, $v_1 = v_s' - v_2'$. Across the shock wave, the flows of the mass, momentum and the energy must be conserved. The conservation equations are

$$\rho_1 v_1 = \rho_2 v_2, \tag{2.130}$$

$$\rho_1 v_1^2 + p_1 + \frac{B_1^2}{2\mu} = \rho_2 v_2^2 + p_2 + \frac{B_2^2}{2\mu}, \tag{2.131}$$

$$\rho_1 v_1 \left( \tfrac{1}{2} v_1^2 + e_1 + \frac{p_1}{\rho_1} \right) + v_1 \frac{B_1^2}{\mu} = \rho_2 v_2 \left( \tfrac{1}{2} v_2^2 + e_2 + \frac{p_2}{\rho_2} \right) + v_2 \frac{B_2^2}{\mu}, \tag{2.132}$$

where $e$ is the internal energy. Here $P$ and $E$ are defined as follows,

$$P = p + \frac{B^2}{2\mu}, \tag{2.133}$$

$$E = e + \frac{B^2}{2\mu}. \tag{2.134}$$

Then eqs (2.131) and (2.132) may be rewritten as

$$\rho_1 v_1^2 + P_1 = \rho_2 v_2^2 + P_2, \tag{2.135}$$

$$\tfrac{1}{2} v_1^2 + \frac{\gamma}{\gamma - 1} \frac{P_1}{\rho_1} = \tfrac{1}{2} v_2^2 + \frac{\gamma}{\gamma - 1} \frac{P_2}{\rho_2}. \tag{2.136}$$

These conservation relations across the shock wave are called the

**Fig. 2.43.** The shock wave and plasma boundary.

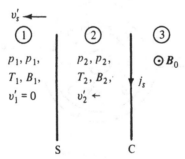

Rankin–Hugoniot relations. From rot $\mathbf{E} = 0$, the electric fields $E = vB$ perpendicular to the magnetic fields are the same within regions ① and ②. That is,

$$v_1 B_1 = v_2 B_2. \tag{2.137}$$

When the values $\rho_1$, $v_1$, $p_1$ and $B_1$ in front of the shock wave are known, the values $\rho_2$, $v_2$, $p_2$ and $B_2$ behind the shock wave can be obtained through eqs (2.130)–(2.132), and (2.137). Equations (2.130)–(2.132) are rewritten to give

$$\frac{\rho_2}{\rho_1} = \frac{v_1}{v_2} = \frac{(\gamma + 1) + (\gamma - 1)P_1/P_2}{(\gamma - 1) + (\gamma + 1)P_1/P_2}, \tag{2.138}$$

$$\frac{T_2}{T_1} = \frac{1}{P_1/P_2} \frac{(\gamma - 1) + (\gamma + 1)P_1/P_2}{(\gamma + 1) + (\gamma - 1)P_1/P_2}, \tag{2.139}$$

as functions of $P_2/P_1$. At the limit $(P_2/P_1) \to \infty$, $(\rho_2/\rho_1) \to (\gamma + 1)/(\gamma - 1)$, which is finite, while $(T_2/T_1) \to \infty$, which is infinite. The shock wave compresses the plasma and in particular heats the plasma greatly. Although the velocity $v_1$ (strictly speaking, the propagation velocity $v_s'$ of the shock wave) is unknown in the region ①, $B_0$ is given by eq. (2.128).

The shock wave collides with itself at the axis of the cylinder and bounces, travelling back into region ②. The expanding shock wave again compresses and heats the plasma.

The disadvantage of the theta-pinch machine is the leaking out of the compressed plasma from the central part of the cylinder toward the two end sections. To compensate for this disadvantage the cylinder has been formed into a torus, around which is a one-turn coil. In the theta-pinch machine, the plasma is compressed and heated unsteadily, and the momentum of the plasma is in balance, sensitively dependent on the dynamics. The curvature of the torus vessel makes the plasma in it unstable, so a theta-pinch machine with a torus shape cannot confine the plasma very well.

### 2.3.9. The z-pinch machine

Through the electrodes set at the two end sections of a cylinder filled with plasma a strong electric current flows along the cylinder axis. A magnetic field in the azimuthal direction is then induced, which compresses and heats the plasma in the cylinder. This kind of experimental device is called a z-pinch machine. The z-pinch machine has been used from the start of fusion experimentation, but in the initial experiments the z-pinch plasma column was soon destroyed by sausage-type or kink-type instabilities.

Recently it has been proposed that a solid DT string, of radius 10 $\mu$m and length 5 cm for example, be set between two electrodes. Even if the input energy is of the order of 1 J and the current in the plasma column of the order of 10 kA, a very strong magnetic field will be induced in the azimuthal direction, confining the plasma because of the small radius of the plasma column. When the plasma density is high (in the solid density) and the radius is small, the z-pinch plasma column is reported to remain stable for several hundred microseconds. There is thus the possibility of achieving the break-even condition using this type of small machine. The main facility required for such a z-pinch machine is a small capacitor bank, so it is very economical, although the output energy per shot cannot be expected to be high.

### 2.3.10. *The plasma focus*

Two co-axial cylinders serve as electrodes, between which the plasma is filled. At the bottom of the co-axial cylinder, radial discharge starts. The radial current sheet $j_r$ induces a magnetic field $B_\theta$ in the azimuthal direction under the current sheet. The Maxwell stress of this magnetic field pushes up the current sheet, and the current sheet makes the plasma flow. After the current sheet arrives at the upper edge of the co-axial cylinder, it separates from the cylinder, curving the surface and forming the closing magnetic surface point 0 in Fig. 2.44, thus confining the compressed and heated plasma. Neutrons released from the plasma due to nuclear fusion reactions can be observed experimentally by the use of this plasma focus.

**Fig. 2.44.** The plasma focus.

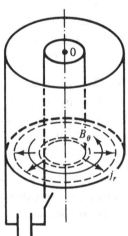

# 3
# The inertial confinement method

•

*This chapter examines the physics that provide the basis of the inertial confinement method, by which nuclear fusion reactions occur when the D–T fuel is heated over a short time interval during which inertia stops the motion of the fuel. Experimental methods for achieving fusion using lasers, the relativistic electron beams, and ion beams are also investigated.*

## 3.1. The fundamentals of the inertial confinement method

### 3.1.1. The Lawson criterion

Dr Bhabha, who was the chief of the Indian Atomic Energy Commission and chairman of the first International Conference for Peaceful Application of Atomic Energy at Geneva in 1955, said in his opening address that, 'mankind will make peaceful use of energy from nuclear fusion within the next 20 years'. Although more than 20 years has passed since then, it will be some time before we utilise fusion energy. Fusion research using the magnetic confinement method has progressed steadily during this period, dealing with some complicated problems, and fusion research using the inertial confinement method is becoming increasingly important, although the history of its research is much shorter than that of the magnetic confinement.

The idea that inertia can be used to confine plasma at a high temperature is quite distinctive, by comparison with the idea that the magnetic field can confine charged particles. The possibility of realising fusion reactions in inertially-confined plasma clearly depends on recent developments in science and technology.

The temperature of the D–T plasma has to be increased to $10^8$ K for both the inertially-confined and the magnetically-confined plasma, with respect to the collision cross-section of the fusion reaction. The Lawson criterion derived in section 1.3.4 for magnetically-confined plasma must now be derived here for inertially-confined plasma.

Using the interial confinement method, the D–T fuel is heated to a temperature of $10^8$ K by an outside supply of energy during the time interval in which the inertia of the fuel itself is effective, and is led to the condition at which fusion reactions break out. Figure 3.1 shows schematically the target for inertial confinement. The D–T fuel is confined in a microballoon, which is made of a thin glass or metal shell. The beam of the driver (the energy source supplied to the target is called the energy driver) deposits energy on the surface layer of the target (in the microballoon); this layer becomes hot, generating high pressure. The surface layer then ablates outward at high speed. As a consequence of this ablation, the layer compresses and heats the fuel. A large part of the beam energy $E_{inp}$ which is absorbed in the target is conveyed as thermal and kinetic energy of plasma which is blown off the target surface. However, part of this energy does work in compressing and heating the D–T fuel. The ratio $\eta_H$ of the energy $E_{DT}$, supplied to the D–T fuel at the instant the fuel is compressed to the highest density, to the total beam energy $E_{inp}$,

$$E_{DT}/E_{inp} = \eta_H \qquad (3.1)$$

is called the hydrodynamic efficiency. The value of $\eta_H$ reaches 10 % when 100 % of the driver energy is absorbed in the target; the target needs to be well made if it is to have a high pellet gain.

Ultimately, the energy $E_{DT}$ is stored in the fuel as thermal energy. Because the initial temperature of the fuel is negligible compared with the final temperature of more than $4 \times 10^7$ K, $E_{DT}$ may be

**Fig. 3.1.** The target and driver beam.

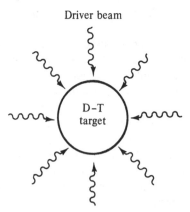

expressed as

$$E_{DT} = \eta_H E_{inp} = 2\left(\frac{4\pi}{3} R^3 \tfrac{3}{2} nkT\right), \qquad (3.2)$$

where $R$ is the final radius of the D–T fuel. In the final stage, the fuel is ionised; the left-hand side of eq. (3.2) is multiplied by 2 for ions and electrons. After the fuel is heated to a high temperature, fusion reactions occur. In the initial state, the fuel satisfies $n_D = n_T = n/2$, where $n$ refers to the number density and the suffixes D and T refer to the deuterium and the tritium, respectively. The reactions then lead to

$$dn/dt = -n^2/2 \cdot \langle \sigma v \rangle. \qquad (3.3)$$

If the average reaction frequency $\langle \sigma v \rangle$ is assumed to be constant (although, in fact, it changes as a function of the temperature), eq. (3.3) gives

$$1/n - 1/n_0 = \tfrac{1}{2}\langle \sigma v \rangle \tau, \qquad (3.4)$$

where $n_0$ is the number density of the fuel before the reaction and $\tau$ is the confinement time of the fuel by inertia (disintegration time of the fuel), expressed by

$$\tau = R/c_s = R/(kT/m_i)^{\frac{1}{2}}, \qquad (3.5)$$

where $c_s = (kT/m_i)^{\frac{1}{2}}$ is the ion sound speed (the expansion velocity of the fuel), $k$ is the Boltzmann constant and $m_i$ is the ion mass of the fuel. If the burning fraction $f$ (the reaction rate) of the fuel is defined by

$$f = (n_0 - n)/n_0, \qquad (3.6)$$

eq. (3.4) can be written as

$$f = \rho R/\{(8m_i c_s/\langle \sigma v \rangle) + \rho R\}, \qquad (3.7)$$

where $\rho = nm_i$ is the density of the fuel. The temperature of the fuel increases to $T = 80$ keV by the self-heating of $\alpha$-particles during the burning. On the average, $T$ is taken to be 20 keV, so that

$$8m_i c_s/\langle \sigma v \rangle \approx 6.3 \text{ g/cm}^2, \qquad (3.8)$$

and eq. (3.7) reduces to

$$f = \rho R/(6.3 + \rho R), \qquad (3.9)$$

where $\rho R$ is expressed in units of g/cm$^2$. When $f$ is expected to have a value of 0.3, $\rho R$ must be 3 g/cm$^2$. In research an inertial confinement fusion, $\rho R$, is frequently used instead of $n\tau$. Between $\rho R$ and $n\tau$ the relation

$$n\tau = (\rho/m_i)(R/c_s) \tag{3.10}$$

holds. The fusion output energy $E_{out}$ from the target is expressed as

$$E_{out} = \tfrac{4}{3}\pi R^3 (n_0 - n) E_f$$
$$= \tfrac{4}{3}\pi R^3 n_0 \rho R E_f / [(8m_i c_s/\langle\sigma v\rangle) + \rho R], \tag{3.11}$$

where $E_f = 17.6$ MeV is the nuclear fusion energy per reaction. The inequality,

$$E_{out} > E_{inp} \tag{3.12}$$

leads to

$$1 < \eta_H \rho R E_f / [2kT\{(8m_i c_s/\langle\sigma v\rangle + \rho R\}], \tag{3.13}$$

by using eqs (3.2) and (3.11). As $T = 4$ keV gives $8m_i c_s/\langle\sigma v\rangle = 40$ g/cm$^2$ ($c_s = 1.2 \times 10^7$ cm/s), eq. (3.13) reduces to

$$\rho R > 2.18 \text{ g/cm}^2, \tag{3.14}$$

when $\eta_H = 0.05$ is chosen. The inequality (3.14) is rewritten as

$$n\tau > 4.2 \times 10^{16} \text{ s/cm}^3, \tag{3.15}$$

by using eq. (3.10). The inequalities (3.14) or (3.15) are the Lawson criterion for the plasma confined by inertia.

### 3.1.2. Compression of fuel

The fusion output energy $E_{out}$ from the target is expected to be more than 3 GJ. If the number of fusion reactions in a target is denoted by $N$, then we have

$$NE_f = E_{out}. \tag{3.16}$$

The mass $M_{DT}$ of the fuel in a target is given by

$$M_{DT} = Nm_i/f = m_i E_{out}/(fE_f). \tag{3.17}$$

If $E_{out} = 3$ GJ and $f = 0.3$, $M_{DT}$ is

$$M_{DT} = 23 \text{ mg}. \tag{3.18}$$

From eq. (3.2), the input energy $E_{inp}$ of the driver is expressed by

$$E_{inp} = 4\pi R^3 nkT/\eta_H. \tag{3.19}$$

If $\eta_H = 0.1$, $T = 4$ keV and $\rho R = 3$ g/cm$^2$, we have

$$E_{inp} = 3.65 \times 10^{12} (n_s/n)^2 \text{ J}. \tag{3.20}$$

Here $n_s$ stands for the number density of the solid fuel and its value is $n_s = 4.5 \times 10^{28}$ m$^{-3}$. When $n = n_s$ we have $E_{inp} = 3.65 \times 10^{12}$ J, which is enormously large and is technologically impossible to supply. However, it is assumed that the fuel can be compressed to $n = 1000 n_s$. Then $E_{inp}$ becomes 3.65 MJ, which is within the range of the energy which can be supplied technologically. When $n = 10^4 n_s$, $E_{inp}$ reduces to 36.5 kJ. In order to achieve the inertial confinement fusion, it is thus necessary to compress the fuel density to $10^3$–$10^4$ times the solid density.

When the fuel is heated to $T = 4$ keV and compressed to $n = 10^{32}$ m$^{-3}$, the pressure $p$ of the fuel has the value $p = 6.4 \times 10^{16}$ Pa, derived from the relation $p = nkT$. For the inertial confinement fusion, the pressure of the fuel must reach high values of the order of $10^{17}$ Pa, which cannot be applied directly on the target by the pusher.

A cryogenic target consisting of three layers has been proposed. The outer layer, which is made of a material of high density (high $Z$), is called the tamper. The tamper contributes to increasing the target gain, stopping the target from being blown out of the material by its heavy mass, and thus decreasing the loss of the thermal and kinetic energy that accompanies the blowing out of the target material. The middle layer, which consists of a material of medium or low density, is called the pusher (Fig. 3.2).

The driver deposits its energy in the pusher layer. The temperature $T_p$ of the pusher layer then reaches $T_p = 100$–$300$ eV, and hence the pressure $p_p$ of the pusher reaches $p_p = 2.4 \times 10^{12}$ Pa. The pusher accelerates the fuel toward the target centre. The mixture of deuterium and tritium becomes solid at a temperature lower than

Fig. 3.2. The cryogenic target.

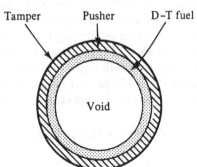

12 K; in the target, which has a temperature of 8 K, the D–T fuel forms a thin solid layer attached to the inside of the pusher layer. For a target radius of $r_t = 6$ mm, for instance, the thickness of the tamper, pusher and fuel layers are each of the order of 100 $\mu$m or less. The target inside the fuel layer is in a vacuum, and the fuel is accelerated toward the centre. Fuel with mass of $M_{DT}$ is accelerated by the pusher layer; the acceleration $\alpha$ is obtained through

$$\alpha M_{DT} = 4\pi r^2 p_p. \tag{3.21}$$

If we choose $r = r_t/2 = 3$ mm, $M_{DT} = 20$ mg and $p_p = 3 \times 10^{12}$ Pa, we have

$$\alpha = 2 \times 10^{13} \text{ m/s}^2. \tag{3.22}$$

The motion of the fuel toward the target centre is called the implosion. Through the relations

$$\tfrac{1}{2}\alpha\tau_{imp}^2 = r_t, \tag{3.23}$$

$$\alpha\tau_{imp} = U, \tag{3.24}$$

the time interval $\tau_{imp}$ of the implosion and the implosion velocity $U$ of the fuel respectively are given by

$$\tau_{imp} \approx 30 \text{ ns}, \qquad U \approx 4.5 \times 10^5 \text{ m/s}. \tag{3.25}$$

During the time interval $\tau_d = 30$ ns of the driver beam, the driver energy of $E_{inp} = 6$ MJ is deposited in the pusher layer in the target. This energy is assumed to be balanced by the work done by the pusher on the fuel. Then we have

$$E_{inp}/\tau_d = 4\pi r^2 p_p U'. \tag{3.26}$$

If we choose $r = r_t/2 = 3$ mm and $U' = U/2 = 2.25 \times 10^5$ m/s, we get

$$p_p = 8 \times 10^{12} \text{ Pa}, \tag{3.27}$$

which is a sufficient value for the pressure of the pusher.

The D–T fuel implodes with a velocity of $U = 5 \times 10^5$ m/s. When the fuel approaches the centre of the target, the surface area of the fuel becomes small and the fuel is decelerated rapidly and stops at the centre through colliding with itself. If we imagine that the fuel is decelerated during $\tau_{dec} = 100$ ps, and the mean radius of the fuel at deceleration is $r_{dec} = 100$ nm, the fuel pressure $p_{DT}$ can be obtained by

$$M_{DT} U = 4\pi r_{dec}^2 p_{DT} \tau_{dec}, \tag{3.28}$$

which gives

$$p_{DT} = 8 \times 10^{17} \text{ Pa}. \tag{3.29}$$

If we choose a fuel temperature of $T = 20$ keV during the nuclear burning and fuel radius $R = 70$ $\mu$m, the confinement time $\tau$ of the fuel is

$$\tau = R/c_s = 600 \text{ ps}. \tag{3.30}$$

We may summarise as follows. D–T fuel of $M_{DT} = 20$ mg is filled in the solid layer in a target of radius $r_t = 6$ mm, which also has a tamper layer. Energy of $E_{inp} = 6$ MJ supplied by a driver whose pulse width is $\tau_d = 30$ ns is deposited in the pusher layer. The pusher is heated to a pressure of $p_p = 6 \times 10^{12}$ Pa and accelerates the fuel with an acceleration of $\alpha = 2 \times 10^{13}$ m/s$^2$ over the time interval $\tau_{imp} = 30$ ns of the implosion. The implosion velocity reaches a value of $U = 5 \times 10^5$ m/s. The fuel is decelerated near the centre; the momentum is converted to an impulse and the deceleration of the fuel induces compression of the fuel, leading to $T = 4$ keV and $p_{DT} = 8 \times 10^{17}$ Pa. The confinement time of the fuel by inertia is $\tau = 600$ ps. During this confinement time, nuclear fusion reactions occur in the fuel and the $\alpha$ particles, a product of the fusion reaction, self-heat the fuel to $T = 80$ keV. The burning fraction reaches $f = 30$ % and the fusion output energy from the target is $E_{out} = 3$ GJ. If we repeat the process with a repetition rate of 1 Hz, we have a fusion reactor whose thermal output is 3 GW.

## 3.2. Absorption of laser light

### 3.2.1. Plasma oscillation

A laser is used at present as the energy driver to heat the D–T plasma to the fusion temperature in the inertial confinement process. In the target in which the D–T fuel is contained, the laser light must be absorbed. Let us investigate in this section the absorption of the laser light in the plasma. The plasma oscillation frequency is an eigenfrequency and plays an important role in the absorption of an electromagnetic wave.

The material which forms the outer shell of the target has a low temperature in the initial stage. However, the energy needed to ionise the material is negligibly small in comparison with the thermal energy in the final state of the material irradiated by the laser light. Therefore the material is assumed to be initially in the state of plasma.

The plasma has a strong tendency to become neutral in charge. In the region where electrons are predominant, these electrons move to neutralise the charge in the electric field which is induced by the dominant charge. By the action of the momenta which electrons obtain from the electric field, electrons pass through a state of equilibrium and form another region dominated by electrons. By repeating this process, electrons are made to oscillate; this oscillation is called plasma oscillation.

To clarify the phenomenon, the situation described in Fig. 3.3 may be examined. If we assume that ions are at rest and only electrons undergo oscillation, we have

$$\varepsilon E = \Sigma,$$

where $\Sigma$ is the charge density on the surface on the left in Fig. 3.3. By using the electron number density $n_e$, $\Sigma$ can be expressed as

$$\Sigma = -n_e ex,$$

where $x$ is the distance of electron displacement. Thus we have

$$E = \frac{-n_e ex}{\varepsilon}. \tag{3.31}$$

The equation of motion for an electron is

$$m_e \frac{d^2 x}{dt^2} = -\frac{n_e e^2 x}{\varepsilon},$$

and electrons are oscillating with a frequency

$$\omega_p = \left(\frac{e^2 n_e}{\varepsilon m_e}\right)^{\frac{1}{2}} = 5.64 \times 10 n_e^{\frac{1}{2}} \text{ rad/s}, \tag{3.32}$$

which is called the plasma frequency. Here $n_e$ is expressed in units of $m^{-3}$.

**Fig. 3.3.** Charge separation in slab plasma.

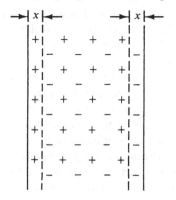

When the following conditions hold for electrons in the plasma,

   (i) ion motions are negligible, and the ion density is uniform and given by $n$;
  (ii) electron thermal motion is negligible;
 (iii) collisions among particles are few;
  (iv) the amplitude of oscillation is small;
   (v) no magnetic field appears;

the equation of continuity and the equation of motion for the electron are respectively

$$\frac{\partial n_e}{\partial t} + \operatorname{div} n_e v_e = 0, \tag{3.33}$$

$$n_e m_e \left( \frac{\partial}{\partial t} + \mathbf{v}_e \cdot \operatorname{grad} \right) \mathbf{v}_e = -n_e e \mathbf{E}. \tag{3.34}$$

In eq. (3.34), the electron pressure, the friction between the electron and the ion, and the magnetic field induced by electron oscillation are neglected. The electron number density is given by

$$n_e(\mathbf{r}, t) = \bar{n}_e + n'_e(\mathbf{r}, t),$$

where $n'_e$ is the perturbed quantity. If terms of second or higher order are neglected, eqs (3.33) and (3.34) reduce to

$$\frac{\partial n'_e}{\partial t} + \bar{n}_e \operatorname{div} \mathbf{v}_e = 0, \tag{3.35}$$

$$\frac{\partial \mathbf{v}_e}{\partial t} = -\frac{e}{m_e} \mathbf{E}. \tag{3.36}$$

The Poisson equation

$$\varepsilon \operatorname{div} \mathbf{E} = e(n_i - n_e)$$

reduces to

$$\operatorname{div} \mathbf{E} = -\frac{e n'_e}{\varepsilon}. \tag{3.37}$$

If $\mathbf{v}_e$ and $\mathbf{E}$ are eliminated by using eqs (3.36) and (3.37), eq. (3.35) leads to

$$\frac{\partial^2 n'_e}{\partial t^2} + \frac{\bar{n}_e e^2}{m_e \varepsilon} n'_e = 0, \tag{3.38}$$

from which it is clear that $n'_e$ oscillates with the frequency given by eq. (3.32).

### 3.2.2. Propagation of electromagnetic waves in plasma

For the case where laser light is radiated onto the plasma, the way
in which the laser light propagates in the plasma will now be studied.
Because laser light is an electromagnetic wave, the electric field **E**
and the magnetic flux density **B** follow the Maxwell equations

$$\operatorname{rot} \mathbf{E} = -\frac{\partial \mathbf{B}}{\partial t}, \tag{3.39}$$

$$\frac{1}{\mu} \operatorname{rot} \mathbf{B} = \mathbf{J} + \varepsilon \frac{\partial \mathbf{E}}{\partial t}. \tag{3.40}$$

The ion motion can be neglected in the plasma, since the ion mass
is much greater than the electron mass. Thus the current density **J**
in the plasma is taken to be constituted by electron motion only.
The equation of motion for the electron is

$$n_e m_e \left( \frac{\partial}{\partial t} + \mathbf{v}_e \operatorname{grad} \right) \mathbf{v}_e = -n_e e(\mathbf{E} + \mathbf{v}_e \times \mathbf{B}), \tag{3.41}$$

where the electron pressure and the friction force caused by the
electron–ion collisions are neglected, following the results of the
preceding section. Neglecting nonlinear terms to do with the electron
velocity $\mathbf{v}_e$ which appear in eq. (3.41), we may simplify eq. (3.41) to

$$\frac{\partial \mathbf{v}_e}{\partial t} = -\frac{e}{m_e} (\mathbf{E} + \mathbf{v}_e \times \mathbf{B}). \tag{3.42}$$

The wave is taken to propagate in the x-direction. The wave number
is denoted by $k_w$ and the frequency is denoted by $\omega$. If all the
dependent variables are expanded in a Fourier series of the form
$e^{i(\omega t - k_w x)}$, then eqs (3.39), (3.40) and (3.42) become

$$ik_w E_y = i\omega B_z, \tag{3.43}$$

$$\frac{ik_w}{\mu} B_z = -en_e v_{ey} + i\omega \varepsilon E_y, \tag{3.44}$$

$$i\omega v_{ey} = -\frac{e}{m_e} (E_y - v_{ex} B_z), \tag{3.45}$$

$$i\omega v_{ex} = -\frac{e}{m_e} v_{ey} B_z. \tag{3.46}$$

In the above equations, the electric field is chosen in the y-direction,
and the magnetic field in the z-direction. The component $E_y$ in

eq. (3.45) is replaced by the $E_y$ of eq. (3.43); thus we have

$$i\omega v_{ey} = -\frac{e}{m_e}(c_p - v_{ex})B_z.$$

On the right-hand side of the above equation, the second term $v_{ex}$, which is the electron velocity in the plasma, is neglected by comparison with the first term $c_p$, which is the speed of light $\omega/k_w$ in the plasma. In other words, the effect of the magnetic field on electron motion is neglected. If we eliminate $v_{ey}$ and $B_z$ in eqs (3.43), (3.44) and (3.45), we get

$$k_w^2 = \frac{1}{c^2}\left(\omega^2 - \frac{n_e e^2}{\varepsilon m_e}\right),$$

where $c^2 = 1/\varepsilon\mu$ is the velocity of light in a vacuum. Since $(n_e e^2/\varepsilon m_e)$ is the plasma frequency $\omega_p$, which was defined by eq. (3.32), the equation described above reduces to

$$k_w^2 = k_0^2\left(1 - \frac{\omega_p^2}{\omega^2}\right), \tag{3.47}$$

where $k_0$ is the wave number of the electromagnetic wave in a vacuum and $\varepsilon_p$, which is defined by

$$\varepsilon_p = 1 - \frac{\omega_p^2}{\omega^2}, \tag{3.48}$$

is the ratio of the dielectric constants. Thus the electric field $E_y$ of the electromagnetic wave in the plasma is described by

$$E_y = E_y^* e^{i(\omega t - k_0\sqrt{\varepsilon_p}x)}, \tag{3.49}$$

where $E_y^*$ is the amplitude of the electric field.

When the number density of the plasma is not great and the plasma frequency $\omega_p$ is lower than the frequency of the electromagnetic wave, the plasma is said to be under-dense. The ratio of the dielectric constants is positive when the plasma is under-dense, as is clear from eq. (3.48). On the other hand, when the number density of the plasma is great and the plasma frequency $\omega_p$ is higher than the wave frequency, the plasma is said to be over-dense. The ratio of the dielectric constants is then negative. When $\varepsilon_p$ is positive, a wave with frequency $\omega$ and the wave number $k_w$ can propagate in the plasma, as eq. (3.49) indicates. On the other hand, when $\varepsilon_p$ is negative, eq. (3.49) is rewritten as

$$E_y = E_y^* e^{-x/\lambda} e^{i\omega t}. \tag{3.50}$$

where

$$\lambda = \frac{1}{k_0(-\varepsilon_p)^{\frac{1}{2}}} = \frac{1}{k_0\left(\dfrac{\omega_p^2}{\omega^2} - 1\right)^{\frac{1}{2}}}, \tag{3.51}$$

Equation (3.50) indicates that the electromagnetic wave is rapidly damped and cannot propagate in the plasma. It can propagate in the under-dense plasma but not in the over-dense plasma.

Lasers which can supply the large amount of energy of 10 kJ during a short time interval of less than 1 ns include the neodimium (Nd) glass laser, which radiates light of wavelength 1.06 $\mu$m, and the carbon dioxide ($CO_2$) laser, which radiates light of wavelength 10.6 $\mu$m. If we choose the Nd laser, which has the shorter wavelength, its frequency is $\omega = 1.8 \times 10^{15}$ rad/s. In the inertial confinement method, the confinement time of the fuel is short. In order to satisfy the Lawson criterion, the density of the fuel must be high, as pointed out in section 3.1.2. If we take a solid fuel whose number density is $n_s = 4.5 \times 10^{28}$ m$^{-3}$, the plasma frequency is $\omega_p = 1.2 \times 10^{16}$ rad/s, given by eq. (3.32). It is quite clear that the fuel is over-dense for the Nd laser (and also of course for the $CO_2$ laser). Accordingly, the light cannot propagate in the plasma and is completely reflected from the plasma surface. This conclusion held up investigation of laser irradiation of plasma until a decade ago, and was the reason why the laser fusion was not a topic of interest until quite recent times.

### 3.2.3. Classical absorption

If solid D–T fuel or another solid material is exposed to a vacuum, molecules evaporate from its surface to give a small vapour pressure. If laser light is irradiated on its surface, the light penetrates the material to a distance of $\lambda$, even though the material is over-dense and light cannot propagate in it. In this layer of thickness $\lambda$ light ionises the material, which subsequently expands. When the main pulse of the laser light arrives at the target, the density distribution of the target resembles that shown in Fig. 3.4. The plasma frequency, which corresponds to the solid density $n_s$, is much higher than the frequency of the light. Near the surface, however, there is a region where the density is low. In this region there is a surface of which the density is $n_c$ and the corresponding plasma frequency is equal to the frequency of the light. The surface $r_c$ where the density is $n_c$ (the critical number density) is called the critical surface. In the region where $r > r_c$, the plasma is under-dense and light propagates to the surface where $r = r_c$. The equation for an electron in the under-dense plasma is

$$\frac{\partial v_e}{\partial t} + \frac{v_e}{\tau} = -\frac{e}{m_e} E, \qquad (3.52)$$

where $\tau$ is the electron mean free flight time with respect to the collisions with ions, and the force due to the magnetic field is neglected. When the electric field of the light oscillates sinusoidally with time, and electron collisions with ions are neglected ($1 \ll \omega\tau$), the electron velocity also varies sinusoidally. Because the phase of the electron velocity is 90 degrees in advance of that of the electric field, the electric field does no work on the electron. However, electrons do collide with ions (of course electrons also collide with one another, but the larger velocity change that occurs is due to collisions with ions), causing the electron to oscillate out of phase. Thus the electric field does work on the electron. The light loses energy, the electrons obtain thermal energy from the light. This process is the inverse of the bremsstrahlung described in section 1.3.2, and is called the inverse bremsstrahlung. The absorption of the light energy through this process is called inverse bremsstrahlung absorption or classical absorption. When the laser light is irradiated on a target which has the density of a solid, classical absorption occurs in the under-dense region where $r > r_c$.

The electric field $E$ and the magnetic field $B$ of the linearly-polarised plane electromagnetic wave propagating in the

Fig. 3.4. Density distribution in the target.

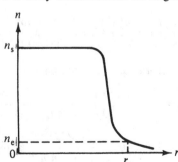

$x$-direction is taken to be oriented in the $y$-direction and the $z$-direction, respectively. Then the Maxwell equations are

$$\frac{\partial E_y}{\partial x} = -\frac{\partial B_z}{\partial t}, \tag{3.53}$$

$$\frac{\partial B_z}{\partial x} = \mu J_y + \frac{1}{c^2}\frac{\partial E_y}{\partial t}, \tag{3.54}$$

where the current $J$ in the $y$-direction is given by

$$J_y = -n_e e v_{ey}. \tag{3.55}$$

All dependent variables are assumed to have the form $\exp i(k_w x - \omega t)$, then, from eqs (3.52)–(3.55), we have

$$\left(\frac{k_w c}{\omega}\right)^2 = 1 - \left(\frac{\omega_p^2}{\omega^2 + 1/\tau^2}\right)\left(1 + i\frac{1}{\omega\tau}\right), \tag{3.56}$$

where $\omega_p$ is the plasma frequency

$$\omega_p^2 \equiv \frac{e^2 n_e}{\varepsilon m}.$$

In general, $k_w$ is complex, and the coefficient $K_v$ of the inverse bremsstrahlung absorption is given by

$$K_v \equiv \mathrm{Im}\{k_w\} = \frac{\omega}{c}\left[-\frac{\beta}{2} + \frac{1}{2}\left\{\beta^2 + (1-\beta)^2\left(\frac{1}{\omega\tau}\right)^2\right\}^{\frac{1}{2}}\right]^{\frac{1}{2}}, \tag{3.57}$$

where

$$\beta \equiv 1 - \frac{\omega_p^2}{\omega^2 + 1/\tau^2},$$

and the frequency $v$ is related by $v = \omega/2\pi$ and the wavelength $\lambda$ is related by $\lambda = c/v$ ($c$ is the velocity of light). When $1/\tau \ll \omega$, eq. (3.56) reduces to

$$k_w \cong \frac{\omega_0}{c}\left(1 - \frac{\omega_p^2}{\omega^2}\right)^{\frac{1}{2}}\left\{1 + i\left(\frac{1}{2\omega\tau}\right)\left(\frac{\omega_p^2}{\omega^2}\right)\frac{1}{1 - n/n_c}\right\}. \tag{3.56'}$$

If we substitute $\tau$ from eq. (2.65), we have

$$K_v = (2\pi\varepsilon^2)^{-1}\left(\frac{1}{3}\right)^{\frac{3}{2}}\frac{Z n_e^2 e^6 \ln \Lambda}{c(m_e k T_e)^{\frac{3}{2}}\omega_0^2(1 - n/n_c)^{\frac{1}{2}}}$$

$$= 2.5 \times 10^{-17}\frac{Z^2\lambda^2}{T^{\frac{3}{2}}(1 - n/n_c)^{\frac{1}{2}}}n^2 \text{ m}^{-1}. \tag{3.57'}$$

The inverse bremsstrahlung absorption of laser light occurs in the under-dense region where $r > r_c$. Classical absorption becomes small with increase in the plasma temperature, while the coefficient $K_v$ becomes large near the critical surface, as shown by eq. (3.57'). But the light energy absorbed in the plasma is finite and small if the absorbed energy is integrated in the under-dense region. When the Nd laser light irradiates a target of the density of a solid, only a small percentage of the light energy is absorbed in the under-dense region of the plasma. Thus by using a Nd laser, it is impossible to achieve inertial confinement fusion solely by the classical absorption.

### 3.2.4. Parametric instability

As the preceding section pointed out, there is no hope of achieving laser fusion if we rely only on classical absorption. This was the main reason why laser fusion research started about ten years later than magnetic-confinement fusion. However, when strong laser light is irradiated on the plane target, more than 90 % of the laser energy is absorbed by the target. A target with the density of a solid is not a perfect reflector, but it is nearly a black body. Thus laser fusion suddenly came into the limelight. On the critical surface, the frequency of the light is equal to the plasma frequency, so the plasma will be in resonance. It is thought that abnormal phenomena occur near the critical surface (Fig. 3.5).

The electromagnetic wave can propagate in the under-dense plasma. If the frequency of the wave is denoted by $\omega$, the wave number in the plasma by $k_w$ and the propagation velocity of the wave in the vacuum by $c$, eq. (3.47) leads to

$$\omega^2 = \omega_p^2 + c^2 k_w^2. \tag{3.58}$$

The ion sound wave (an electrostatic wave) propagates in the plasma. The propagation velocity $c_s$ of the ion sound wave is given by

**Fig. 3.5.** Irradiation of the plane target by laser light.

Plane target

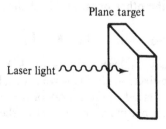

Laser light

$c_s = (kT_e/m_i)^{\frac{1}{2}}$, which was derived in section 3.1.1. If the frequency of the ion sound wave is denoted by $\omega$ and its wave number by $k_w$, we obtain

$$\omega = k_w c_s = k_w \left(\frac{kT_e}{m_i}\right)^{\frac{1}{2}}. \tag{3.59}$$

When the electron temperature is much higher than the ion temperature $T_i$ in the plasma, the ion sound wave can propagate in the plasma without strong damping.

The plasma has an eigenfrequency $\omega_p$ (which is the plasma frequency and was described in section 3.2.1). In section 3.2.1, the plasma frequency was written for the purely stationary oscillation which does not propagate, because the temperature of the plasma was neglected entirely. If we take the thermal velocity $v_{et}$ of the electron into consideration, the frequency $\omega$ is related to the wave number $k_w$ by

$$\omega^2 = \omega_p^2 + \frac{k_w^2}{2} v_{et}^2. \tag{3.60}$$

This wave propagates in the plasma and is called the electron plasma wave, or the Bohm–Gross wave. The electron plasma wave is damped when its wave number is large and $k_w r_D$ is of O(1), where $r_D$ is the Debye radius of the plasma. The electron plasma wave can propagate in the plasma without a strong damping if $k_w r_D < 1$.

When a strong monochromatic electromagnetic wave (laser light) propagates in the plasma, a wave such as the ion sound wave or the electron plasma wave sometimes grows, excited by some plasma parameter, for example the density or the temperature, which specifies the plasma properties. These kind of phenomena are called parametric instabilities. Decay instability occurs most frequently among the parametric instabilities when, for example, the three waves described above interact with each other in the plasma. If the frequencies and the wave numbers of the three waves are denoted respectively by $(\omega_1, \mathbf{k}_1)$, $(\omega_2, \mathbf{k}_2)$ and $(\omega_3, \mathbf{k}_3)$, the energy of one wave is transferred to the other ones, provided that the resonance conditions

$$\omega_1 = \omega_2 + \omega_3, \qquad \mathbf{k}_1 = \mathbf{k}_2 + \mathbf{k}_3, \tag{3.61}$$

are satisfied. If we specify the laser light as $(\omega_1, \mathbf{k}_1)$, the ion sound wave as $(\omega_2, \mathbf{k}_2)$ and the electron plasma wave as $(\omega_3, \mathbf{k}_3)$, the energy of the laser light is transferred to the energy of the eigenwaves, ion sound wave and electron plasma wave subject to the conditions of eq. (3.61). This suggests a new process of absorption of the laser light. If we draw the three waves in the $\omega$–$k_w$ plane, the conditions of (3.61) become clear. In Fig. 3.6, curve 1 shows the wave for eq. (3.58), curve 2 the wave for eq. (3.60), and curve 3 the wave for eq. (3.59). When the three vectors **A**, **B** and **C**, which start from the origin and terminate on the points on curves 1, 2 and 3, satisfy

$$\mathbf{A} = \mathbf{B} + \mathbf{C},$$

then the three waves represented by these three vectors satisfy the resonance condition (3.61). The parametric instability that causes the laser light to decay to the ion sound wave and the electron plasma wave causes the abnormal absorption of the laser light. This has led to laser fusion becoming the subject of research. In addition to decay instability, there are three other parametric instabilities: Raman scattering, by which laser light decays into another electromagnetic wave and an electron plasma wave in the region where $\omega_1 > 2\omega_p$; Brillouin scattering, by which laser light decays into another electromagnetic wave and an ion sound wave in the region where $\omega_1 > \omega_p$; and finally two-plasma instability, by which laser light decays into two electron plasma waves in the region where $\omega_1 > 2\omega_p$.

Parametric instability can be explained by the ponderomotive force. The wave number of the electromagnetic wave approaches zero near the critical surface, as shown by eq. (3.47). Therefore the wavelength there approaches infinity. If we neglect the thermal motion and the friction due to collisions with ions, the equation of

**Fig. 3.6.** Condition for decay instability in three waves.

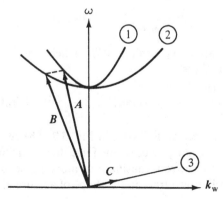

motion for the electron is

$$m_e \frac{\partial \mathbf{v}}{\partial t} = -e(\mathbf{E} + \mathbf{v} \times \mathbf{B}). \qquad (3.62)$$

Because the wave number $k_w$ can be considered to be zero near the critical surface, the electric field of the wave there is described by

$$\mathbf{E}(\mathbf{r}, t) = \mathbf{E}^*(\mathbf{r}) \cos \omega_1 t, \qquad (3.63)$$

oscillating with the frequency $\omega_1$. The amplitude $E^*(\mathbf{r})$ is a function of $\mathbf{r}$. From one of the Maxwell equations $\partial \mathbf{B}/\partial t = -\text{rot } \mathbf{E}$, we deduce

$$\mathbf{B}(\mathbf{r}) = -\frac{1}{\omega_1} \text{rot } \mathbf{E}^*(\mathbf{r}) \sin \omega_1 t. \qquad (3.64)$$

The second term on the right-hand side of eq. (3.62) is small in comparison with the first term, which was described in section 3.2.2. The first approximate solution of eq. (3.62) is

$$\mathbf{v}^{(1)} = -\frac{e}{m_e \omega_1} \mathbf{E}^*(\mathbf{r}_0) \sin \omega_1 t = -\mathbf{v}^* \sin \omega_1 t,$$
$$\left( \mathbf{v}^* = \frac{e \mathbf{E}^*(\mathbf{r}_0)}{m_e \omega_1} \right). \qquad (3.65)$$

The modification for $\mathbf{v}^{(1)}$ comes from the fact that $\mathbf{E}(\mathbf{r})$ is a function of $\mathbf{r}$. The electron position in the absence of the electric field is indicated by $\mathbf{r}_0$. The displacement of the electron from $\mathbf{r}_0$ by the action of the electric field is denoted by $\delta\mathbf{r}$. Then $\mathbf{E}(\mathbf{r})$ is expanded to

$$\mathbf{E}(\mathbf{r}) = \mathbf{E}(\mathbf{r}_0) + (\delta\mathbf{r} \cdot \text{grad})\mathbf{E}(\mathbf{r}_0), \qquad (3.66)$$

and $\delta\mathbf{r}$ is obtained as

$$\delta\mathbf{r} = \frac{e}{m_e \omega_1^2} \mathbf{E}^*(\mathbf{r}_0) \cos \omega_1 t. \qquad (3.67)$$

Thus the equation governing $\mathbf{v}^{(2)}$, which modifies the first approximation $\mathbf{v}^{(1)}$, is

$$\frac{m_e \partial \mathbf{v}^{(2)}}{\partial t} = -e\left[ (\delta\mathbf{r} \cdot \text{grad})\mathbf{E}(\mathbf{r}_0) + \frac{1}{c} \mathbf{v}^{(1)} \times \mathbf{B}(\mathbf{r}_0) \right]. \qquad (3.68)$$

The trace of the electron is shown in Fig. 3.7. The second term on the right-hand side of eq. (3.68) is the force due to the magnetic field which acts on the electron, causing it to move perpendicular to the electric field and draw a trace in the shape of a figure 8. The average during one period is denoted by $\langle \ \rangle$. If we define $\mathbf{f}_{NL}$ by

$$\mathbf{f}_{NL} = m_e \left\langle \frac{\partial \mathbf{v}^{(2)}}{\partial t} \right\rangle, \tag{3.69}$$

$\mathbf{f}_{NL}$ is given by

$$\mathbf{f}_{NL} = -\frac{e^2}{m_e \omega_1^2} \tfrac{1}{2} [\mathbf{E}^*(\mathbf{r}_0) \cdot \text{grad}) \mathbf{E}^*(\mathbf{r}_0) + \mathbf{E}^*(\mathbf{r}_0) \times \text{rot } \mathbf{E}^*(\mathbf{r}_0)], \tag{3.70}$$

which is the low-frequency (steady) force originating from the nonlinear coupling of the electromagnetic wave with high frequency $\omega_1$. Since $\mathbf{f}_{NL}$ acts on an electron, the nonlinear force $\mathbf{F}_{NL}$ per unit volume is given by $\mathbf{F}_{NL} = n_e \mathbf{f}_{NL}$, which is called the ponderomotive force. With eq. (3.70), $\mathbf{F}_{NL}$ is given by

$$\mathbf{F}_{NL} = -\frac{\omega_p^2}{\omega_1^2} \nabla \frac{\varepsilon \mathbf{E}^{*2}}{4}$$

$$= -\frac{\omega_p^2}{\omega_1^2} \nabla \frac{\varepsilon \langle \mathbf{E}^2 \rangle}{2}. \tag{3.71}$$

As Fig. 3.8 shows, the $x$-axis is chosen to be in the direction of $\mathbf{E}^*$. Let us suppose that a perturbation $n'$ of a number density whose wave number if $k_3$ is induced along the $x$-axis. At the instant when the electric field is toward the positive $x$-direction, the electron moves in the negative $x$-direction. Accordingly, the charges are separated due to the density perturbation $n'$, and the local electric field $\mathbf{E}'$ appears in the $x$-direction, as in Fig. 3.8(b). Equation (3.71) is transformed to give

$$\frac{2}{\varepsilon} \frac{\omega_1^2}{\omega_p^2} \mathbf{F}_{NL} = -\nabla \langle \mathbf{E} + \mathbf{E}' \rangle^2 \approx -2(\mathbf{E}^* \cdot \text{grad}) \mathbf{E}' = 2\mathbf{E}^* \frac{\partial \mathbf{E}'}{\partial x}.$$

Thus the ponderomotive force $\mathbf{F}_{NL}$ acts in the directions shown in Fig. 3.8(c), and the density perturbation grows.

**Fig. 3.7.** Trace of the electron oscillation, excited by an electromagnetic wave.

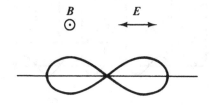

On the critical surface, the wave number $k_w$ of the electromagnetic wave becomes zero and the frequency $\omega_1$ of the wave is equal to the plasma frequency due to eq. (3.58). On the other hand, the frequency $\omega_2$ of the electron plasma wave is given by eq. (3.60), and if the wave number of the electron plasma wave is denoted by $k_2$, then $\omega_1 < \omega_2$. When the inequality $\omega_1 < \omega_2$ holds and the two wave frequencies are close enough to each other, the density perturbation described above is readily induced. This perturbation grows but does not propagate. This kind of instability is called an absolute instability, while the absolute instability shown in Fig. 3.8 is called the oscillative two-stream instability.

When an electron whose eigenfrequency $\omega_2$ executes a forced oscillation due to the external electric field $E^* \cos \omega_1 t$, the displacement $x$ of the electron is governed by

$$\frac{d^2 x}{dt^2} + \omega_2^2 x = -\frac{e}{m_e} E^* \cos \omega_1 t,$$

whose solution is given by

$$x = -\frac{e}{m_e} E^* \frac{\cos \omega_1 t}{\omega_2^2 - \omega_1^2}.$$

**Fig. 3.8.** Density perturbation and ponderomotive force.

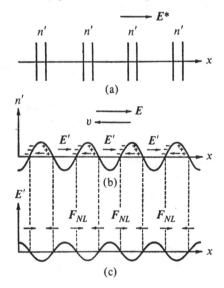

The displacement occurs in the opposite direction to the electric field when $\omega_1 < \omega_2$, and the oscillative two-stream instability grows. In the region where the density is a little less than the critical density, we have $\omega_1 > \omega_2$, and the electron moves the same direction as that of the electric field. As a result, the separated charge induces a perturbation $E_0'$ in the electric field. The ponderomotive force due to $E'$ reduces the density perturbation, as Fig. 3.9 shows. Thus the absolute instability occurs only in the case where $\omega_1 < \omega_2$. What happens in the case where the density perturbation $n'$ is caused by the ion sound wave (wave frequency $\omega_3$ and wave number $k_3$), which propagates along the propagation direction of the electromagnetic wave of $(\omega_1, k_1)$ with the velocity $c_s$? If we observe the phenomenon in a frame of reference which moves with the ion sound wave, the frequency of the electromagnetic wave has a Doppler shift of

$$\omega_1' = \omega_1 - \omega_3 = \omega_1 - k_3 c_s.$$

When the inequality $\omega_1' < \omega_2$ is satisfied, the density perturbation $n'$ again grows, i.e. the energy of the electromagnetic wave excites and amplifies the ion sound wave via the electron plasma wave. This is called decay instability.

Although the frictional force caused by electron–ion collisions is neglected in the above description, in reality this frictional force suppresses the electron wave motion and damps the perturbation. The ponderomotive force must outweigh the frictional force if it is to make the perturbation grow. The lowest value of the intensity of the electromagnetic field for which the perturbation grows is called the threshold. The threshold of the oscillative two-stream instability is higher than that of the decay instability, except in the case where the electron temperature is extremely high. Therefore decay

**Fig. 3.9.** Density perturbation and ponderomotive force for $\omega_1 > \omega_2$.

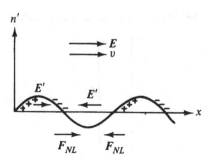

instability is the first to be induced near the critical surface, and is the usual parametric instability. This instability leads to the abnormal absorption of the laser light. The threshold varies for different wavelengths of the laser; it is $10^{16}$ W/m$^2$ for a Nd laser of the wavelength of 1.06 m. The threshold of the $CO_2$ laser is two orders of magnitude less than that of the Nd laser.

At the end of this section we shall investigate backward scattering instability. When laser light propagates in the under-dense region, it is thought that the electrostatic wave $(\omega_2, k_2)$ is induced in the direction of the laser propagation, and another electromagnetic wave $(\omega_3, k_3)$ is induced in the opposite direction to the laser propagation. We assume here that the three wave numbers satisfy

$$\mathbf{k}_1 = \mathbf{k}_2 + \mathbf{k}_3,$$

and the three frequencies satisfy

$$\omega_1 = \omega_2 + \omega_3.$$

If the electric field of the laser light is written as

$$\mathbf{E}_1 = \mathbf{E}_1^* \cos(\omega_1 - k_1 y),$$

then the electron velocity $\mathbf{v}_1$ can be written as

$$\mathbf{v}_1 = -\frac{e\mathbf{E}_1^*}{m_e \omega_1} \sin(\omega_1 t - k_1 y),$$

where $\mathbf{v}_1$ advances with a phase shift of 90 degrees to $\mathbf{E}_1$. The ponderomotive force $\mathbf{F}_{NL}$ caused by the two waves of $\omega_1$ and $\omega_3$ is

$$\mathbf{F}_{NL} = -\frac{n_e e^2}{m_e \omega_1^2} \langle (\mathbf{E}_1 \cdot \mathrm{grad})\mathbf{E}_3 + (\mathbf{E}_3 \cdot \mathrm{grad})\mathbf{E}_1 \rangle$$

$$- e n_e \langle \mathbf{v}_1 \times \mathbf{B}_3 + \mathbf{v}_3 \times \mathbf{B}_1 \rangle.$$

**Fig. 3.10.** Wave vectors in backward scattering.

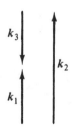

If we choose the $x$-direction to be in the direction of $\mathbf{E}_1$, the gradient is in the $y$-direction. Thus the terms to do with the electric field in $\mathbf{F}_{NL}$ cancel each other, while the terms to do with the magnetic field remain.

Figure 3.11 shows the fields at the instant when the maximum value of $\mathbf{B}$ overlaps with the maximum value of $\mathbf{v}_3$. The direction of $\mathbf{B}_1$ is in the same direction as $\mathbf{k}_1 \times \mathbf{E}_1$ and perpendicular to the plane of the page. The ponderomotive force caused by $-\mathbf{v}_3 \times \mathbf{B}_1$ acts downward. We next observe the instant when the maximum value of $\mathbf{B}_1$ is just opposite the maximum value of $\mathbf{v}_3$ after they advance one-quarter of a wavelength. The ponderomotive force acts upward, as shown in Fig. 3.12. During one period, the ponderomotive force alternates with every quarter of a wavelength.

**Fig. 3.11.** The case where the maximum $\mathbf{B}_1$ coincides with the maximum $\mathbf{v}_3$.

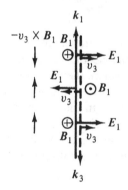

**Fig. 3.12.** The case where the maximum of $\mathbf{B}_1$ is in the opposite direction to that of $\mathbf{v}_3$.

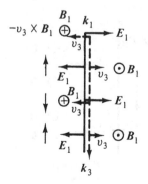

Thus the ponderomotive force induces the density increments $n_2'$ at every half of the wavelength, as shown in Fig. 3.13. These density increments induce the currents $\mathbf{j}' = -n_2'e\mathbf{v}_1$ in the $x$-direction at every wavelength. Finally, current induces a magnetic field $\mathbf{B}'$ which is in phase with $\mathbf{B}_3$ and enhances it.

This parametric instability is called stimulated Raman scattering if $(\omega_2, \mathbf{k}_2)$ is the electron plasma wave, and stimulated Brillouin scattering if $(\omega_2, \mathbf{k}_2)$ is the ion sound wave. In both the cases, the electromagnetic wave $(\omega_3, \mathbf{k}_3)$ scatters backward when the laser light propagates. The thresholds of both backward scatterings are higher than that of the decay instability.

### 3.2.5. Resonance absorption

Let us suppose here that the plasma is located in the region of positive $x$, and that the number density $n$ is proportional to $x$ (Fig. 3.14). When at a distance $x = L$ the number density $n = n_c$, then

$$n = \frac{n_c x}{L}. \tag{3.72}$$

When laser light radiates parallel to $x$, the electric field is perpendicular to $x$ and the electron oscillates on the equi-density surface. Consider the case where the laser is radiated on the plasma obliquely; the $x$–$y$ plane is chosen to include the incident light. When the electric field of the light is in the $x$–$y$ plane, the light is said to be P-polarised; when the electric field is in the $z$-direction, the light is said to be S-polarised. The electric field of the P-polarised light contains the $x$-component and is entirely in the $x$-direction at the

**Fig. 3.13.** Density perturbation induced by the ponderomotive force.

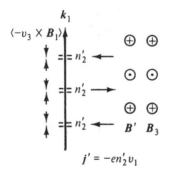

turning point A. Thus the electron oscillates along the density gradient, and it is possible to induce a strong wave due to the charge separation.

When the laser light radiates perpendicular to the plasma surface, the light can penetrate to the critical surface Q. However, light whose incident angle is $\theta$ advances to A according to geometrical optics, as in Fig. 3.15, and not to Q. When the density gradient is large ($L$ is small) and the intensity of the laser light is strong, light permeates to Q, according to physical optics. On the surface Q resonance occurs, inducing a wave of a large amplitude because the light frequency is equal to the plasma frequency.

To investigate the proparation of light in the plasma, let us take the Maxwell equations and the equation of motion for the electron,

$$\text{rot } \mathbf{E} = -\frac{\partial \mathbf{B}}{\partial t}, \qquad (3.73)$$

**Fig. 3.14.** Density profile of the plasma.

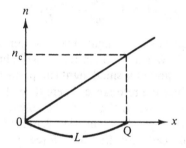

**Fig. 3.15.** Oblique incidence of P-polarised light.

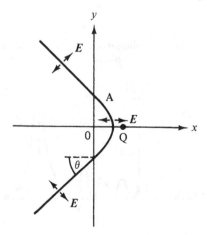

$$\frac{1}{\mu} \, \text{rot} \, \mathbf{B} = \mathbf{J} - \varepsilon \frac{\partial \mathbf{E}}{\partial t}, \qquad (3.74)$$

$$\frac{\partial \mathbf{v}}{\partial t} = -\frac{e}{m_e} \mathbf{E}, \qquad (3.75)$$

$$\mathbf{J} = -ne\mathbf{v}. \qquad (3.76)$$

The light is assumed to be P-polarised; $\mathbf{B} = (0, 0, B_z)$, $\mathbf{E} = (E_x, E_y, 0)$ and the dependent variables have the form $\exp i(\omega t - k_y y)$ with respect to $t$ and $y$. If we eliminate $B_z$, $E_y$, $\mathbf{v}$, and $\mathbf{J}$ in eqs (3.73)–(3.76) we obtain an equation for $E_x$ as follows:

$$\frac{d^2 E_x}{dx^2} + k_0(\varepsilon_p - \sin \theta)E_x + \frac{d}{dx}\left(E_x \frac{d \log \varepsilon_p}{dx}\right) = 0, \qquad (3.77)$$

where $k_0$ is the wave number in the vacuum, and $\varepsilon_p$ is the ratio of dielectric constants given by

$$\varepsilon_p = 1 - \frac{\omega_p^2}{\omega^2} = 1 - \frac{x}{L}, \qquad (3.78)$$

and gives the frequency of the light. The solution $E_x$ of eq. (3.77) is shown in Fig. 3.16. When the intensity of the light is high and the distance between A and Q is small, light can permeate to the critical surface Q and induce the resonance. There the value of $E_x$ diverges. On the basis of $E_x$ shown in Fig. 3.16, the absorption coefficient $A$ of light in the plasma is a function of $(k_0 L)^{\frac{1}{3}} \sin^2 \theta$, and is given in Fig. 3.17. (In this case, the frictional force due to electron–ion collisions is totally neglected. The energy of the light is absorbed only on the surface Q where $E_x$ is infinite.) As Fig. 3.17 shows, $A$

Fig. 3.16. Intensity of obliquely incident P-polarised light in plasma.

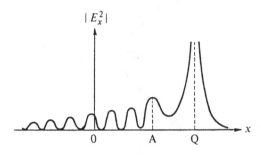

reaches 40 % under the most desirable circumstance. Since the electron has a thermal motion, the equation of motion (3.75) for the electron must include a pressure term, which leads to the finite (but large) value of $E_x$ at Q. From the ion the electron accepts the frictional force, which increases the absorption of light. Here we assume that $n$ is linearly proportional to $x$, from eq. (3.72). In reality, the ponderomotive force of the laser light modifies the density profile in front of Q, which enhances absorption. Sometimes the coefficient $A$ of the absorption reaches 90 %. This kind of absorbing process is called resonance absorption.

The mechanisms of parametric absorption and resonant absorption have now been clarified for laser light which radiates the plasma. A target with over-dense plasma does not reflect 100 % of the incident laser light, but it is a high absorber of that light. Thus laser fusion has become a major research topic. Recently high-power lasers have revealed that absorption increases stepwise with the intensity of the incident light, due to the parametric and resonance absorptions.

The frequency is $\omega = 1.8 \times 10^{15}$ rad/s and the critical density is $n = 8.5 \times 10^{26}/\text{m}^3$ for a Nd laser of wavelength 1.06 $\mu$m. If we assume that the temperature at the critical surface is $T = 1\,\text{keV} = 10^7$ K, the pressure $p_c$ there is

$$p_c = nkT = 10^6 \text{ atm}. \tag{3.79}$$

When the laser, whose energy is 1 kJ, irradiates a target whose spot radius is 100 $\mu$m, the intensity of the laser on the target surface is $10^{20}$ W/m$^2$ and the photon pressure $3 \times 10^6$ atm. Since the photon pressure exceeds the plasma pressure on the critical surface, the

**Fig. 3.17.** The absorption coefficient in resonance absorption.

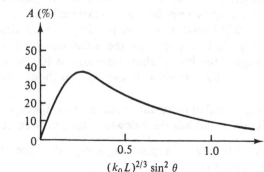

photon pressure modifies the density profile in the under-dense region. The density gradient becomes steep and $L$ becomes small. Thus resonance absorption becomes more effective. Resonance absorption occurs above $10^{18}$ W/m$^2$ for Nd laser light, and above $10^{15}$ W/m$^2$ for $CO_2$ laser light.

### 3.2.6. Laser light of short wavelength

The absorption rate of the laser light is clearly increased by resonance absorption. On the other hand, the resonance phenomenon induces a strong electric field at the critical surface. This field propagates from the high-density side to the low-density side with a phase velocity. The energy of the laser light is absorbed in the plasma by the resonance absorption. This absorbed energy, however, is not converted to electron thermal energy there, but rather is converted to kinetic energy of a small number of electrons, called suprathermal electrons, which are ejected from the target with high velocities, being accelerated by the resonance field. The suprathermal electrons are pulled back to the target by the induced electrostatic field and pass to and fro around the target surface. The amplitudes of these oscillations of the suprathermal electrons are large; the electrons penetrate into the deep part of the fuel to preheat it. The rate of conversion of energy from the laser to the suprathermal electrons is higher when the wavelength of the laser is long. This rate of conversion is nearly 100 % for the $CO_2$ laser. The suprathermal electrons do not compress the fuel but do preheat it. Given this fact, resonance absorption offers the possibility of playing a negative role in extracting nuclear fusion energy. A sophisticated target must be devised, one which has a structure able to convert the kinetic energy of the suprathermal electrons into thermal energy on the target surface. The most common view in recent times is that the $CO_2$ laser's wavelength is too long to serve as a driver for fusion.

The KrF laser is an excimer laser which extracts ultra-violet light (wavelength is 0.249 $\mu$m). The corresponding critical density is $n_c = 1.8 \times 10^{28}$ m$^{-3}$, which is near the solid density $n_s = 4.5 \times 10^{28}$ m$^{-3}$. Energy of this kind of short-wave laser is 100 % absorbed in the target by the classical absorption, without abnormal absorptions.

A large single crystal of potassium phosphate ($KH_2PO_4$) is called KDP. This KDP crystal has the following excellent properties:

  (i) it is possible to grow a giant single crystal whose radius is several tens of cm;

(ii) the conversion rate of the wavelength is high;

(iii) minimal damage occurs when intense laser light passes through it.

When Nd laser light (1.06 $\mu$m) passes through the KDP crystal, the original light is converted to the second harmonic (0.53 $\mu$m) when the resonance condition $\omega_1 = \omega_2 + \omega_3$ is satisfied ($\omega_2 = 2\omega_1$). We choose the crystal as follows. After the original light passes through the crystal, the ratio of the intensity of $\omega_1$, to that of $\omega_2$ is unity. After the combination of the light of $\omega_1$ and $\omega_2$ has passed through the second crystal, we obtain the third harmonic $\omega_3$ (ultra-violet, 0.35 $\mu$m) under the resonance condition $\omega_3 = \omega_1 + \omega_2 = \omega_1 + 2\omega_1$. By using the Nd laser as a driver in recent inertial fusion research experiments, the third harmonic of the Nd laser light radiates on the target and the energy is absorbed classically.

The laser light in the ultra-violet band has the disadvantage that optical components such as mirrors and lenses tend to be damaged by it. No nonlinear optical crystal like KDP has been discovered for laser light of longer wavelength than 10.6 $\mu$m.

## 3.3. Implosion of the target

### 3.3.1. The hydrodynamics of implosion

With the discovery of abnormal absorption, laser fusion opened the first gate leading to its goal. Such a qualitative jump is always necessary in the development of the science and technology. The recent exploitation of new lasers with short wavelengths and techniques of converting the frequency of the laser to the higher harmonics are also important in developing the field of laser fusion. It is necessary to compress the fuel to a high density $10^3$–$10^4$ times that of the density of solids, during a time interval of the order of 10 ns in order to achieve inertial-confinement fusion, as explained in section 3.1.2. To compress the fuel a pressure of $10^{12}$ Pa $= 10^7$ atm is required. It is impossible to compress the fuel using only photon pressure of laser light.

The possibility of implosion of the target is suggested by the following hydrodynamic analysis (Fig. 3.18). The absorbed energy near the critical surface in the target irradiated by the laser light is converted to electron thermal energy (actually, to kinetic energy of the suprathermal electrons). If the laser has a short wavelength, electrons in the under-dense region $r > r_c$ are heated by classical absorption. That is, the temperature, and hence the pressure, of the electrons near the critical density or in the under-dense region increases. The pressure gradient expels electrons outward. The electric field, outwardly directed, is induced by the charge separation because the mobility of the ion is low. Thus the plasma, including ions, undergoes ambipolar diffusion via this electric field. In reaction to this outward motion of the plasma in the corona region, compression waves induced by the pressure gradient propagate inward to the plasma. The compression waves overlap to form shock waves. Section 2.3.5 described the shock wave propagating toward the target centre. In inertial-confinement fusion, the magnetic field is not applied to the target. Thus we let $\mathbf{B} \to 0$ in eqs (2.130)–(2.132). As described in section 2.3.5, the temperature of the plasma behind the shock wave increases without limit, but the ratio of the densities before and after the shock wave approaches $\rho_1/\rho_2 \to (\gamma + 1)/(\gamma - 1)$ when the intensity $p_2/p_1$ of the shock wave becomes high. The ratio $\gamma$ is $\frac{5}{3}$ for the plasma. Thus $\rho_1/\rho_2 \to 4$; i.e. the density behind the shock wave increases at best to four times the density in front of the shock wave. To achieve fusion, an increase in temperature of the plasma after the shock wave is preferable. However, the density of the plasma must also be increased to a high density, $10^3$–$10^4$ times the solid density. If a strong shock wave heats the plasma to a high temperature, but only compresses the plasma to four times the solid density, it is difficult to effect further compression of the plasma because of its high pressure. To obtain highly compressed fuel, continuous compression and heating of the fuel plasma (adiabatic compression) by successive weak shock waves (from sound waves) is required.

Implosion of the target may be analysed using the hydrodynamic equations. The Debye radius of the plasma in the target is given by

**Fig. 3.18.** Density and pressure profile in the target.

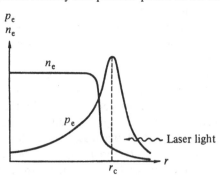

$r_D = 0.1\,\mu$m, if we substitute $T_e = 10^3$ K, $n = 10^{30}$ m$^{-3}$ (average number density of the fuel) into eq. (2.57). Since the Debye radius is so small, the charge separation in the target plasma can be neglected, and the electrons and ions can be considered to move together. We can therefore use the equation of continuity and the equation of motion for the plasma, including the electrons and ions. On the other hand, the mean free flight time $\tau$ of the electron with respect to collisions with ions is of the order of $\tau = 10$ ps, according to eq. (2.65). As $\tau$ increases with $T^{\frac{3}{2}}$, $\tau$ approaches the confinement time of the fuel when the electron temperature becomes high. In this sense, collisions between the electron and the ion are not expected to be frequent. In other words, energy transfer between electrons and ions is not effective. The difference may appear in the electron and ion temperatures. For this reason, we choose the two energy equations for the electron and the ion, respectively. If we assume that the target implodes in a spherically symmetric way, the equation of continuity is

$$\frac{\partial \rho}{\partial t} + \frac{1}{r^2}\frac{\partial \rho u r^2}{\partial r} = 0. \tag{3.80}$$

The equation of motion is

$$\frac{\partial u}{\partial t} + u\frac{\partial u}{\partial r} = -\frac{1}{\rho}\frac{\partial p}{\partial r} - \frac{4}{3\rho r^2}\frac{\partial}{\partial r}\left(r^2\mu\frac{\partial u}{\partial r}\right) - \frac{4u}{3\rho}\frac{\partial}{\partial r}\left(\frac{\mu}{r}\right) - \frac{4\mu u}{3\rho r^2}. \tag{3.81}$$

In these equations, $\rho$ is the density, $u$ the radial velocity, $p$ the pressure, and $\mu$ the coefficient of viscosity. When fusion reactions occur strongly, we must add sink terms to both equations. The equation of energy for the ion is

$$\frac{\partial T_i}{\partial t} + u\frac{\partial T_i}{\partial r} = -\frac{2m_i p_i}{3\rho k r^2}\frac{\partial r^2 u}{\partial r} + \frac{T_e - T_i}{\tau} + W_i + Q_i, \tag{3.82}$$

where $T_i$ is the ion temperature. The first term on the right-hand side of eq. (3.82) gives the work done by the ion pressure $p_i$, and the second term gives the rate of energy transfer from the electron to the ion caused by collisions. The third term is

$$W_i = f E_\alpha N, \tag{3.83}$$

where $N$ is the frequency of the nuclear reaction (number of reactions per unit time and per unit volume), and $E_\alpha$ is the energy of an $\alpha$-particle produced by a nuclear reaction. $E_\alpha = 3.5$ MeV for the D–T reaction. Since the neutrons produced by nuclear reactions fly out from the target without colliding with plasma particles, we discard the energy of the neutron in eq. (3.82). (If $\rho R \approx 5$ g/cm$^2$, neutrons deposit a part of their energy in the target; if $\rho R \approx 7$ g/cm$^2$, the rate of the energy deposition in the target reaches 50 %.) The rate $f$ given by

$$f = (1 + 32/T_e)^{-1} \qquad (T_e \text{ in keV})$$

of the α-particle is absorbed by the ion and the rate $1 - f$ is absorbed by the electron. As derived in eq. (1.21), $N$ is given by

$$N = \tfrac{1}{4}\langle \sigma v \rangle n^2. \tag{3.84}$$

In the plasma, the α-particle has a finite stopping range. If we wish to observe the α-particle more carefully, we must add another energy equation for that particle. In eq. (3.82), α-particles are assumed to be absorbed instantaneously at the places where they are produced. The last term in eq. (3.82) gives the dissipation of kinetic energy of plasma with coefficient of the ion viscosity

$$Q_i = \frac{8\mu_i}{9kn}\left(\frac{\partial u}{\partial r}\right)^2 - \frac{16 u \mu_i}{9knr}\frac{\partial u}{\partial r} + \frac{8\mu_i}{9kn}\left(\frac{u}{r}\right)^2. \tag{3.85}$$

The energy equation for the electron is

$$\frac{\partial T_e}{\partial t} + u\frac{\partial T_e}{\partial r} = -\frac{2m_i p_e}{3\rho k r^2}\frac{\partial r^2 u}{\partial r} - \frac{2m_i c}{3\rho k r^2}\frac{\partial r^2 I}{\partial r} - \frac{T_e - T_i}{\tau}$$

$$+ \frac{2m_i}{3\rho k r^2}\frac{\partial}{\partial r}\left(r^2 K_e \frac{\partial T_e}{\partial r}\right) + W_e + Q_e - A\rho T_e^{\frac{1}{2}}. \tag{3.86}$$

The second, fourth and seventh terms on the right-hand side in the above equation have no corresponding terms in eq. (3.82) for the ion. The second term gives the energy absorbed from the laser light. In this equation, $c$ is the speed of light and $I$ is the energy flux density of the laser light. For $I$, we have

$$\frac{1}{r^2}\frac{\partial r^2 I}{\partial r} + K_a I = 0, \tag{3.87}$$

where $K_a$ is the coefficient of absorption of laser light in the plasma, and is determined by the value for classical absorption in the under-dense region and the value for abnormal absorption near the critical surface. The fourth term comes from the thermal flux, and $K_e$ is the coefficient of electron thermal conductivity. The corresponding term is discarded in eq. (3.82) because the coefficient of ion thermal conductivity is rather smaller than that of the electron.

The seventh term gives the energy loss by bremsstrahlung, which is specified by eq. (1.44) (divided by $\rho$ for the unit mass). The viscous dissipation term $Q_e$ is given if we replace the suffix i by e in eq. (3.85). The energy $Q_e$ by the fusion reaction is given by

$$W_e = (1 - f)E_\alpha N. \tag{3.88}$$

The motion of the target may be analysed numerically using a computer, given the initial conditions for the target and with $I$ at $r = \infty$ a function of time. It is possible to compress the fuel to $10^3$–$10^4$ times the solid density and so to obtain sufficient fusion energy from the target.

### 3.3.2. Rayleigh–Taylor instability

The density profile of the plasma in the imploding target is drawn schematically in Fig. 3.19. The surface at $r = r_c$ is the discontinuous surface at which the density jumps to the critical density value; it is called the deflagration wave (or combustion wave). The energy of the laser light is absorbed at this surface and the deflagration wave sends shock waves propagating to the target centre. In the figure, two shock waves are shown, together with the three discontinuous surfaces at $r_1$, $r_2$ and $r_c$.

When oil is poured into a vessel first and then heavier water is poured onto the oil, as in Fig. 3.20, the situation is unstable because of gravitational force. The fluctuations appearing on the boundary between the oil and the water grow rapidly and eventually the water changes place with the oil. This kind of instability is called Rayleigh–Taylor instability.

**Fig. 3.19.** Surface of the density jump in the imploding target.

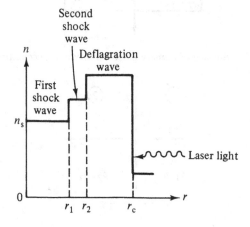

There are three surfaces of the density discontinuity in Fig. 3.19: a deflagration wave and two shock waves. When the deflagration wave moves to the target centre with an inward acceleration, the d'Alembert force acts on it from the inside to the outside. The situation is similar to that shown in Fig. 3.20, and the deflagration wave will manifest Rayleigh–Taylor instability. Also, when the shock wave propagates to the centre with a deceleration, the shock wave too will manifest Rayleigh–Taylor instability. It is necessary that such Rayleigh–Taylor instability does not grow significantly before the implosion motion of the target has ended, if there is to be successful realisation of inertial-confinement fusion.

Let us investigate here the growth rate of the Rayleigh–Taylor instability. As Fig. 3.21 shows, the fluid flows across the surface $D$ of the density discontinuity under the acceleration $\mathbf{g}$. The region in front of the surface is denoted by ①, the region after the surface by ②. The equations of continuity, motion and energy for the fluid are

$$\frac{\partial \rho}{\partial t} + \operatorname{div} \rho \mathbf{v} = 0,$$

$$\rho \left\{ \frac{\partial \mathbf{v}}{\partial t} + (\mathbf{v} \cdot \operatorname{grad}) \mathbf{v} \right\} = -\operatorname{grad} p + \rho \mathbf{g},$$

**Fig. 3.20.** The Rayleigh–Taylor instability.

Water

Oil

**Fig. 3.21.** Surface of density discontinuity and acceleration.

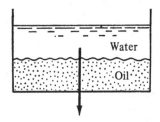

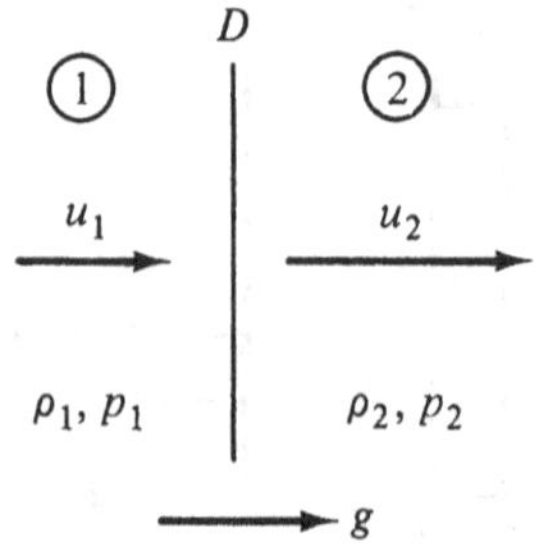

$$\frac{\partial}{\partial t}(\rho e + \tfrac{1}{2}\rho v^2) + \operatorname{div}\{\mathbf{v}(\rho e + \tfrac{1}{2}\rho v^2 + p)\} = \rho\mathbf{v}\cdot\mathbf{g} + I_a\delta(\mathbf{r} - \xi).$$

In the equation of energy, $e$ stands for the internal energy and $I_a$ stands for the laser energy absorbed at the discontinuous surface $D$. The rate of the absorbed energy $I_a$ per unit time and per unit area is assumed here to be constant, and to be absorbed at $\mathbf{r} = \xi$. The $x$-direction is chosen in the direction of the undisturbed flow velocity. If we integrate the above equations with respect to $x$ over the distance from the point just in front of the surface to the point just behind the surface, Rankine–Hugoniot relations similar to eqs (2.130)–(2.132) can be obtained.

With regard to the disturbance, the surface is assumed to be displaced a distance $x = \xi(y)$ from $x = 0$ where the undisturbed surface is located. The perturbations appear in both regions ① and ②, and the dependent variables are changed to $p_1 + p_1'$, $u_1 + u_1'$, $v_1'$, $p_2 + p_2'$, and $v_2'$, where the prime refers to the perturbed quantities. The velocity component $v$ is in the $y$-direction. For simplicity, the fluid is assumed to be incompressible in both regions, and density fluctuations in both regions are neglected. All perturbed quantities are assumed to have the form $\exp(\omega t + iky)$ with respect to $t$ and $y$, and the conservation equations across the discontinuous surface are reformulated as follows to include the perturbations:

The equation of continuity is

$$-\gamma\xi(\rho_1 - \rho_2) + \rho_1 u_1^* - \rho_2 u_2^* = 0, \tag{3.89}$$

the equation of motion in the $x$-direction is

$$p_1^* - p_2^* + 2\rho_1 u_1 u_1^* - 2\rho_2 u_2 u_2^* = -\xi g(\rho_1 - \rho_2), \tag{3.90}$$

and the equation of energy is

$$\gamma\xi\rho_1 u_1(u_1 - u_2) + (\tfrac{5}{2}p_1^* u_1 - \tfrac{5}{2}p_2^* u_2 + \tfrac{5}{2}p_1 u_1^* - \tfrac{5}{2}p_2 u_2^*$$
$$- \tfrac{3}{2}\rho_1 u_1^2 u_1^* - \tfrac{3}{2}\rho_2 u_2^2 u_2^*) = 0. \tag{3.91}$$

The quantities with an asterisk refer to values of perturbations (indicated by primes) at $x = 0$ where the position of surface $D$ is approximated linearly.

The equation of continuity, the equation of motion in the $x$-direction and the equation of motion in the $y$-direction for the perturbations in the region ① are

$$\frac{\partial u_1'}{\partial x} + ikv_1' = 0, \tag{3.92}$$

$$\rho_1\left(\gamma u_1' + u_1 \frac{\partial u_1'}{\partial x}\right) = -\frac{\partial p_1'}{\partial x}, \tag{3.93}$$

$$\rho_1\left(\gamma v_1' + u_1 \frac{\partial v_1'}{\partial x}\right) = -\mathrm{i}k p_1'. \tag{3.94}$$

If we eliminate $v_1'$ and $p_1'$ from these three equations, the equation for $u_1'$ can be solved to give

$$u_1' = u_1^* \mathrm{e}^{\pm kx}. \tag{3.95}$$

In region ①, the perturbations tend to zero as $x$ tends to $-\infty$. Thus we must choose

$$u_1' = u_1^* \mathrm{e}^{kx}. \tag{3.96}$$

If we substitute $u_1'$ from eq. (3.96) into eqs (3.92) and (3.94), we obtain

$$u_1^* + \mathrm{i}v_1^* = 0, \tag{3.97}$$

$$\rho_1 v_1^*(\gamma - ku_1) = -\mathrm{i}k p_1^*. \tag{3.98}$$

The equations governing the perturbations in region ② are obtained by replacing the suffix 1 in eqs (3.92)–(3.94) by suffix 2. The perturbations must tend to zero when $x$ tends to $\infty$. Thus instead of eq. (3.96), we have

$$u_2' = u_2^* \mathrm{e}^{-kx}. \tag{3.99}$$

The equations corresponding to eqs (3.97) and (3.98) are

$$u_2^* - \mathrm{i}v_2^* = 0, \tag{3.100}$$

$$\rho_2 v_2^*(\gamma - ku_2) = -\mathrm{i}k p_2^*. \tag{3.101}$$

For the seven unknowns $\xi$, $u_1^*$, $v_1^*$, $p_1^*$, $u_2^*$, $v_2^*$ and $p_2^*$, we now have the seven equations (3.89), (3.90), (3.91), (3.97), (3.98), (3.100) and (3.101). In order to obtain non-zero solutions for these seven homogeneous equations, the determinant consisting of the coefficients of the unknown variables in the seven equations must be zero. That is,

$$\begin{vmatrix} -\gamma(\rho_1 - \rho_2) & \rho_1 & -\rho_2 & 0 & 0 & 0 & 0 \\ g(\rho_1 - \rho_2) & 2\rho_1 u_1 & -2\rho_2 u_2 & 0 & 0 & 1 & -1 \\ \gamma\rho_1 u_1(u_1 - u_2) & \frac{5}{2}p_1 + \frac{3}{2}\rho_1 u_1^2 & -\frac{5}{2}p_2 - \frac{3}{2}\rho_2 u_2^2 & 0 & 0 & \frac{5}{2}u_1 & -\frac{5}{2}u_2 \\ 0 & 1 & 0 & \mathrm{i} & 0 & 0 & 0 \\ 0 & 0 & 0 & \rho_1(\gamma + ku_1) & 0 & \mathrm{i}k & 0 \\ 0 & 0 & 1 & 0 & -\mathrm{i} & 0 & 0 \\ 0 & 0 & 0 & 0 & \rho_2(\gamma - ku_2) & 0 & \mathrm{i}k \end{vmatrix} = 0.$$

$$\tag{3.102}$$

If we change the variables into dimensionless forms with

$$\zeta = \frac{\gamma}{(kg)^{\frac{1}{2}}}, \qquad U = \frac{kv_2}{(kg)^{\frac{1}{2}}}, \qquad Q = \frac{p_1}{\rho_2 u_2^2}, \qquad \varepsilon = \frac{\rho_2}{\rho_1}, \quad (3.103)$$

the dimensionless growth rate $\zeta$ must satisfy

$$5\zeta^2 U - \zeta^2 U(13 + 5Q) - 5\zeta U(1 + \varepsilon U^2) + U^2(13 + 5Q) = 0. \quad (3.104)$$

In the case in which $u_1 = u_2 = 0$ (the case where the flow does not cross the surface D (Fig. 3.22)), the well-known growth rate of the classical Rayleigh–Taylor instability

$$\gamma = (kg)^{\frac{1}{2}} \qquad (3.105)$$

is obtained from eq. (3.104). The growth rate given by eq. (3.105) is proportional to $k^{\frac{1}{2}}$. But in practice perturbations with high wave numbers dissipate their energies at rates determined by transport coefficients such as the viscosity and thermal conductivity, and the growth rate decreases. When the thickness of the layer of the density jump is taken into account, the growth rate of the Rayleigh–Taylor instability is obtained as the eigenvalue of the boundary problem. Usually the existence of a finite density gradient causes the growth rate of the instability for the discontinuous surface to be decreasing.

## 3.4. Lasers

### 3.4.1. Inverse population and stimulated emission

Let us now investigate the laser. An atom (or molecule or ion) is assumed to have three atomic energy levels. In Fig. 3.23, ① shows

**Fig. 3.22.** A fluctuation appearing on the surface of a density discontinuity.

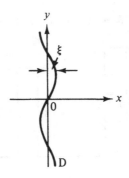

the basic state. Two excited states whose energy levels are separated by only a very small gap are denoted by ② and ③. $W_{12}$ and $W_{13}$ are the absorption transition probabilities from level ① to level ② and to level ③ respectively, absorbing light energy in the process. $A_{12}$ and $A_{13}$ are the natural transition probabilities, by which an atom at level ② and at level ③ respectively transits to the basic state ①, this time emitting light. The probabilities $\bar{W}_{21}$ and $\bar{W}_{31}$ denote the induced transition probabilities by which an atom at level ② and at the level ③ respectively transits to basic state ①, induced by the corresponding light energies which pass through the atom. The notation $S_{32}$ indicates the nonradiation transition probability by which an atom transits from energy level ③ to energy level ② without the emission of the light. If the numbers of atoms at level ①, ② and ③ are referred to as $N_1$, $N_2$ and $N_3$,

$$\frac{dN_3}{dt} = W_{13}N_1 - (\bar{W}_{31} + A_{31} + S_{32})N_3, \qquad (3.106)$$

$$\frac{dN_2}{dt} = W_{12}N_1 - (\bar{W}_{21} + A_{21})N_2 + S_{32}N_3, \qquad (3.107)$$

may be derived. Since the total number of atoms is constant, we have

$$N_1 + N_2 + N_3 = N_0. \qquad (3.108)$$

In the steady state, $dN_2/dt = dN_3/dt = 0$, eqs (3.106) and (3.107) lead to

$$\frac{N_2}{N_1} = \left[ \frac{S_{32}\bar{W}_{31}}{\bar{W}_{31} + A_{31} + S_{32}} + \bar{W}_{21} \right] \Big/ [A_{21} + \bar{W}_{21}].$$

When the nonradiation transition probability is so large that the inequalities $A_{31} \ll S_{32}$, $\bar{W}_{31} \ll S_{32}$ are satisfied, the above equation

**Fig. 3.23.** A three energy level laser.

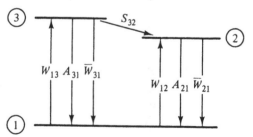

is reduced to

$$\frac{N_2 - N_1}{N_0} \approx \frac{W_{13} - A_{21}}{W_{13} + A_{21} + 2W_{12}}. \tag{3.109}$$

If $W_{13} > A_{21}$ holds, we have $N_2 > N_1$. This inequality shows that the number $N_2$ of atoms in the higher level ② is greater than the number $N_1$ in the lower level ①. In this situation it is said that the atoms have an inverse population. In order to emit laser light, the atoms must have such an inverse population. If the frequency of the light corresponding to the energy discrepancy between level ② and level ① is denoted by $v$, the equality

$$E_{12} = hv$$

holds, where $h$ is the Planck constant. Light of frequency $v$ and flux intensity $I_v$ is thought to propagate in a medium which includes the atoms under consideration. If the increase in the intensity of the light after that light has propagated a distance $dx$ is denoted by $dI_v$, we have

$$dI_v = N_2\, dx\rho_v \bar{B}_{21}hv - N_1\, dx\rho_v B_{12}hv \quad \left( \bar{B}_{21} = \frac{\bar{W}_{21}}{\rho_v}, B_{12} = \frac{W_{12}}{\rho_v} \right). \tag{3.110}$$

In eq. (3.110), the term to do with the natural transition probability $A_{21}$ is omitted because $A_{21}$ is much smaller than the induced transition probability $\bar{W}_{21}$. If the energy density of the light is denoted by $\rho_v$, then

$$\rho_v = \frac{I_v \dfrac{dx}{c}}{dx} = \frac{I_v}{c}. \tag{3.111}$$

If $\rho_v$ is eliminated in eq. (3.110), we obtain

$$dI_v = hv(\bar{B}_{21}N_2 - B_{12}N_1)I_v \frac{dx}{c}. \tag{3.112}$$

**Fig. 3.24.** Change in the flux intensity of light.

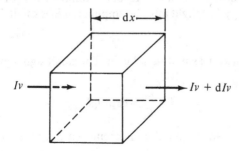

The intensity $I_v$ of the light flux after propagation along a distance $x$ is given by

$$I_v = I_{v_0} e^{-K_v x},$$

where $I_{v_0}$ is the initial intensity of the light flux and $K_v$ is the absorption coefficient of the medium. The differential form of the above equation is

$$\frac{dI_v}{dx} = -K_v I_v. \tag{3.113}$$

If the term $dI_v/dx$ from eq. (3.113) is substituted into eq. (3.112), then $K_v$ can be expressed as

$$K_v = \frac{h\nu}{c}(B_{12}N_1 - \bar{B}_{21}N_2). \tag{3.114}$$

With $g_1$ and $g_2$ indicating the degeneracies of levels ① and ②, the absorption and the induced probabilities $W_{12}$ and $\bar{W}_{21}$ respectively may be expressed by

$$B_{12} = \frac{c^3 A_{21}}{8\pi h\nu^3}\frac{g_2}{g_1}, \qquad \bar{B}_{21} = \frac{c^3 A_{21}}{8\pi h\nu^3}.$$

Thus eq. (3.114) leads to

$$K_v = \frac{h\nu}{c}\frac{c^3 A_{21}}{8\pi h\nu^3}\frac{g_2}{g_1}\left(N_1 - \frac{g_1}{g_2}N_2\right). \tag{3.115}$$

When the atoms have the inverse population $N_1 < N_2$ and the inequality $N_1 < g_1 N_2/g_2$ is satisfied, $K_v$ is negative. In such a situation, the intensity of the light increases after it propagates in the medium. This kind of stimulated emission, with one phase and one wavelength, is called a laser.

A high-power laser system consists of an oscillator and several amplifier stages, which determine the character and net efficiency of the system. The output energy $E_o$ of an amplifier is proportional to the input energy $E_i$ of light into the amplifier, if $E_i$ is small. That is,

$$E_o = GE_i, \tag{3.116}$$

where $G$ is called the small signal gain. The small signal gain is given by

$$G = \frac{E_o}{E_i} = \exp(g_0 L), \tag{3.117}$$

where $g_0$ is the small signal gain coefficient and $L$ is the effective length of the amplifier. The output energy $E_o$ is not proportional to $E_i$ when $E_i$ becomes large because the stored energy density in the laser medium is finite. The gain for a strong input energy with a pulse shaped in double steps is called the large signal gain and is given by

$$G = \frac{E_o}{E_i} = \frac{E_s}{E_i} \ln\left[ 1 + \left\{\exp\frac{E_i}{E_s} - 1\right\} \exp(g_0 L)\right] \quad (3.118)$$

where $E_s$ is the saturated energy density in the medium. When $E_i \gg E_s$, eq. (3.118) gives

$$E_o - E_i = E_s g_0 L. \quad (3.119)$$

The atoms in the amplifier must maintain the inverse population until the input pulse reaches them after they have been excited. In order to amplify the laser intensity, the value of $g_0$ for the medium needs to be large. But if $g_0$ is too large automatic emission occurs and the medium cannot accumulate energy. The gain coefficient $g_0$ is given by

$$g_0 = \sigma \Delta N,$$

where $\sigma$ is the cross-sectional area of the induced transition and $\Delta N$ is the number density of the inverse population. When $\sigma$ is too large, $g_0$ becomes finite even for small $\Delta N$. The cross-sectional area $\sigma$ of the induced transition is given by

$$\sigma = \frac{A\lambda_0^2}{8\pi^2 \Delta v}, \quad (3.120)$$

where $A$ is the natural transition probability, $\lambda_0$ is the mean wavelength and $\Delta v$ is the full width of the half maximum (FWHM) of the transition spectrum. In order to extract laser light with high power and a short pulse, a medium with a small $A$ and a large $\Delta v$ must be prepared. The parameters for the Nd glass laser, $CO_2$ laser, I laser, HF laser and KrF laser are given in Table 3.1.

### 3.4.2. The neodymium glass laser

Figure 3.25 shows the oscillator of an Nd glass laser. The xenon lamp winding around the glass rod flashes upon discharge of the electric energy stored in the high voltage capacitor bank. The $Nd^{3+}$ ion is excited to the $^4F_{3/2}$ level by the xenon light. The induced transition of $Nd^{3+}$ from the $^4F_{3/2}$ level to the $^4I_{1/2}$ level emits light of wavelength 10.6 $\mu$m. One edge surface of the glass rod (oscillator) is a mirror, while the other edge is a half mirror (made of condensed

Table 3.1 *Parameters of high-power lasers*

| Laser | Wavelength (cm) | Natural transition probability $A$ (s$^{-1}$) | FWHM of spectrum (Hz) | Cross-sectional area of the induced transition (cm$^2$) | Small signal gain coefficient $g_0$ (cm$^{-1}$) | Saturated energy density $E_s$ (J/cm$^2$) | Total stored energy (J/l) |
|---|---|---|---|---|---|---|---|
| Nd glass | 1.06 | $5 \times 10^3$ | $3 \times 10^{12}$ | $2.5 \times 10^{-20}$ | 0.15 | | 500 |
| $CO_2$ | 10.6 | 0.84 | $2 \times 10^{10}$ (1 atm) | $1.0 \times 10^{-18}$ (1 atm) | 0.05 | $0.01 \sim 0.02$ | 15 (3 atm) |
| I | 1.315 | 8 | | $2 \times 10^{-18}$ ($CO_2$, 1 atm) | 0.08* | 0.6 | 30 |
| HF | $2.6 \sim 3.0$ | | | $6.5 \times 10^{-17}$ | | | 100 (6 lines) |
| KrF | 0.248 | | $2 \times 10^{12}$ | $\sim 10^{-17}$ | 0.05 | | |

* $C_3F_7I$: 40 Torr, $CO_2$: 240 Torr

metal vapour). The intensity of the light increases rapidly in the oscillator, excited ions effecting the induced transition. When the intensity of the light between the mirrors becomes very great, the laser light passes through the half mirror and is emitted.

To protect the glass of the amplifier from damage, the intensity of laser light must be limited. If the total energy of laser light is to be high, the radius of the glass rod must be large. However, for a rod of greater than 5 cm radius, cooling inside the rod is not effective and the possibility of thermal damage to the rod is considerable. The amplifier in the final stage has disc glasses, as shown in Fig. 3.26. The radius of the light which passes through the final amplifier may sometimes reach 20 cm. On the other hand, the driver energy must be directed at the target over 10 ns, which is the effective time interval of the target implosion. It is possible to shorten the pulse width of the Nd laser to 100 ps, during which the light advances 3 cm. The laser light at the instant of arrival of the light on the

**Fig. 3.25.** Oscillator for laser light.

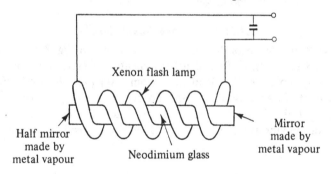

Xenon flash lamp

Half mirror
made by
metal vapour

Neodimium glass

Mirror
made by
metal vapour

**Fig. 3.26.** Disc-type amplifier for laser light.

High-voltage
DC source

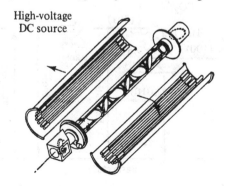

target chamber is in the shape of a disc of radius 20 cm and height 3 cm. The light is focused by a lens onto the target, whose radius is of the order of 100 $\mu$m (experimentally). If the total energy of the light is 1 kJ, the pulse width is 100 ps, and the surface area of the target $10^{-8}$ m$^2$, then the intensity of the light on the target surface is $10^{21}$ W/m$^2$. The maximum value of the electric field at this intensity of light reaches $10^{12}$ V/m, which accelerates electrons to the velocity of light over one period.

The Nd glass laser is well equipped by present technology to extract high-power light over a short time interval. However, it has only low efficiency, about 0.1 % with respect to the energy conversion from electricity to light, which makes it difficult to extract a large amount of laser energy, say 10 MJ, at a low price. Since the Nd laser is a solid laser, cooling is not effective. It is necessary to wait several hours before the temperature of the rod returns to its initial value after just one shot. As a result, repetitive operation cannot be expected. The Nd glass laser is valuable in demonstrating target implosion but cannot serve as a real driver for extracting practical fusion energy from the target.

### 3.4.3. The carbon dioxide laser

The simplified levels of the $CO_2$ laser are shown in Fig. 3.27. In the figure (001), (100), (020) and (010) indicate the vibrational modes of a $CO_2$ molecule. The transition from level (001) to level (100) emits light of wavelength 10.6 $\mu$m, and from level (001) to level (020) emits light of 9.6 $\mu$m. However, a detailed description of the energy levels of the vibrational modes of $CO_2$ molecules will not be given.

**Fig. 3.27.** Vibrational energy levels for the $CO_2$ laser.

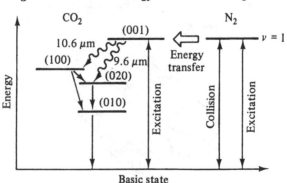

The $CO_2$ laser has many spectral lines between 9–18 $\mu$m; the two lines which have the strongest spectra are chosen here.

A mixture of the gases $CO_2$ and $N_2$ is used in the $CO_2$ laser. The $CO_2$ molecules are excited by inelastic collisions with electrons and generate an inverse population. The energy level of vibration $v = 1$ of the $N_2$ molecules has an almost equivalent energy level to level (001) of the $CO_2$ molecule. (The lifetime of the level of $v = 1$ of the $N_2$ molecule is long. Discharge in $N_2$ gas induces many $N_2$ molecules at level $v = 1$.) If the superscript ↑ refers to the excited state, the $CO_2^\uparrow$ appears as eq. (3.121):

$$CO_2 + e \rightarrow CO_2^\uparrow + e, \qquad (3.121)$$

or

$$N_2 + e \rightarrow N_2^\uparrow + e, \qquad (3.122)$$

$$N_2^\uparrow + CO_2 \rightarrow CO_2^\uparrow + N_2. \qquad (3.123)$$

The rate of energy conversion of the $CO_2$ laser is high for a continuous wave but low for a pulsed wave whose pulse width is of order of a few ns. Every vibrational level consists of many rotational levels, as shown in Fig. 3.28. The transitions between the rotational levels follow the selection rule. Imagine that the light is amplified in the $CO_2$ laser by using the transition from level $J = 19$ in (001) to level $J = 20$ in (100) (shown by P(20) in Fig. 3.28). When the pulse width is long, the molecules at the other levels in (001) move to $J = 19$ (relaxation) in the case where the number of molecules at level $J = 19$ decreases by the transition. However, if the pulse width is very short, the molecules at level of $J = 19$ only contribute to amplifying the light. Methods of improving the efficiency of the

**Fig. 3.28.** Rotational energy levels for the $CO_2$ laser.

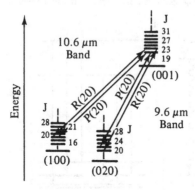

pulsed $CO_2$ laser are:

   (i) with a high gas pressure of the laser, the relaxations between
       the rotational levels become short;
  (ii) multi-lines in addition to the line of P(20) can be used for
       the amplification of the light, so too multi-bands for 10.6 $\mu$m
       and for 9.6 $\mu$m.

The $CO_2$ laser has the following advantages:

   (i) the energy conversion efficiency is in the range 2–5 %;
  (ii) construction of a high-power system is quite easy.

On the other hand, the $CO_2$ laser has the fatal disadvantage that
the wavelength is too long (10.6 $\mu$m) to compress the fuel in the
target.

### 3.4.4. The iodine laser

The I laser emits light of wavelength 1.315 $\mu$m by using the magnetic
dipole transition from $^2P_{1/2}$ to $^2P_{2/3}$. The energy level diagram for
the I laser is given in Fig. 3.29. As the laser media, $CF_3I$ and $C_3F_7I$
are used, absorbing light energy of wavelength 2500–2900 Å and
dissociating to the excited atom $I^\dagger$ ($^2P_{1/2}$). On the other hand, the
basic state $I(^2P_{3/2})$ is produced only slightly by the photo-
dissociation. Thus population inversion is generated. The reaction
equation is

$$C_3F_7I + h\nu \rightarrow I^\dagger + C_3F_7. \qquad (3.124)$$

The characteristics of the I laser are:

   (i) The wavelength of 1.315 $\mu$m is long (ultra-red), but is much
       shorter than that of the $CO_2$ laser.

**Fig. 3.29.** Energy levels in the I laser.

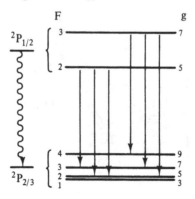

(ii) It is rather easy to highly pressurise the gas, to construct a large system, and to provide a short pulse.

(iii) Although the transition probability is small, since the transition is in the magnetic dipole type, the width of spectra is also small. If we add another gas to the gas medium and raise the gas pressure, the width of spectra becomes large and the energy stored in the medium increases.

(iv) The efficiency of the laser is at most 0.5–1 %, limited by the efficiency of excitation by the flash lamp. To increase this low efficiency, a new method of excitation needs to be invented.

(v) For repetitive operation, the laser medium must be reproduced via the reaction ($C_3F_7 + I \to C_3F_7I$). The energy required for such reproduction of the laser medium decreases the efficiency of the laser.

### 3.4.5. The hydrogen fluoride laser

The HF laser uses the following reactions:

$$F + H_2 \to HF^\dagger + H \qquad -\Delta H = 31.6 \text{ kcal/mole}, \qquad (3.125)$$

$$H + F_2 \to HF^\dagger + F \qquad -\Delta H = 98 \text{ kcal/mole}. \qquad (3.126)$$

Using the above reactions, the excited HF molecule is produced: by eq. (3.126), excited molecules with $v \leqslant 3$ are produced; by eq. (3.127), molecules with $v \leqslant 6$ are produced. Figure 3.30 shows the vibration levels for the reaction given by eq. (3.126) and the population rates of the energy levels. Population inversion occurs

**Fig. 3.30.** Vibration levels in the HF laser.

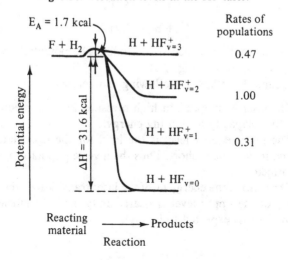

first for levels $v = 2$ and $v = 1$, and the laser emission starts. Subsequently the transitions $v = 3 \rightarrow v = 2$ and $v = 1 \rightarrow v = 0$ follow.

An electron beam is used to excite the HF laser. This electron beam separates the F atom from the F molecule or the F compounds, and triggers the reactions of eq. (3.126); but the laser does not excite the HF molecule directly as in the case of the $CO_2$ laser. Thus the two efficiencies, electrical and chemical, refer to the HF laser. The characteristics of the HF laser are:

   (i) The wavelength is between 2.7–3.0 $\mu$m, which is a little too long.
  (ii) Because the reaction consumes a large amount of H and F, the electrical efficiency alone is not sufficient. The energy needed to reproduce the H and F must be taken into account. Net efficiency is expected to be 3–5 %.
 (iii) Ballistic oscillation (self-emission of light) occurs due to the large cross-sectional area of the induced transition; the laser is not of the energy-storage type, although it is expected to be of high power.
 (iv) It is difficult to make the pulse width short. The system cannot be fine tuned to extract a homogeneous beam.
  (v) Techniques of extracting light without risk of accident are required.

### 3.4.6. The excimer laser

When an excited atom $A^\dagger$ forms a loose coupling with either a similar or a different atom, i.e.

$$A^\dagger + A \rightarrow (A_2)^\dagger, \tag{3.127}$$

$$A^\dagger + B \rightarrow (AB)^\dagger, \tag{3.128}$$

then $(A_2)^\dagger$ or $(AB)^\dagger$ is called the excimer. The potential energy of the excimer is shown in Fig. 3.31.

The excimer laser has the following characteristics:

   (i) The laser is operated at high pressure, and the medium has a high capacity for storing energy.
  (ii) The lower energy level is the level for the dissociated gas, and its lifetime is short. Thus the inverse population is easily formed.
 (iii) The ratio of the energy $h\nu$ of the laser transition to the energy $h\nu_p$ of the upper level is nearly unity, and the efficiency of the laser is expected to be high.

(iv) The transition occurs with a wide spectrum band and the laser can easily be made to give short pulses.

The excimers of rare gases such as Xe, $Ar_2$ and $Kr_2$, and the excimers of rare-gas haloids such as XeCl, XeF, XeB and KrF, are now being investigated. All of them have some disadvantages and none is a definite candidate to serve as the driver of the fusion. However, KrF, which has a short wavelength of 0.249 μm (ultra-violet), is one of the most likely drivers for laser fusion.

### 3.4.7. The free-electron laser

In the lasers described above, the atom (or molecule or ion) is designed to form the inverse population using a flash lamp, an electric discharge, or an electron beam. When the atom is excited using these methods, the conversion of electrical energy to the energy of the atoms in the excited level is not expected to be high. Unfortunately, the efficiency of this kind of laser is generally too low for a driver in laser fusion, unless a target with a very high target gain can be developed in the future; currently it is impossible to produce a target with a target gain of more than 1000.

The free-electron laser has a quite different emission mechanism. If the electron beam is projected into a periodic magnet, the electron paths are curved by the action of the magnetic field. The electrons emit electromagnetic waves according to their acceleration, as described in section 1.3.2. When the wavelength of the ripples in the magnetic field generated by a Whigler coil is $\lambda_w$ and the speed of an electron before entering the coil is $v_e$ (usually equal to the velocity

**Fig. 3.31.** Potential curves of the excimer laser.

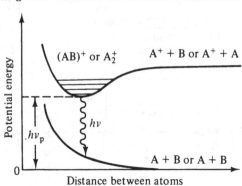

of light $c$), the wave frequency $\omega$ of the emitted laser light is given by

$$\omega = 2\pi v_e / \lambda_w. \tag{3.130}$$

If the kinetic energy of an electron emerging from the periodic magnet is recirculated (the electron is accelerated again to its original speed and is returned to the periodic magnet), the efficiency is expected to be more than 25 %. (The efficiency of the extraction of the electron beam is about 50 %, as described in section 3.4.1.) From now on, much of the research on the free-electron laser will concentrate on the fact that in such a laser the intensity and the beam energy can be increased, and the pulse width can be made short.

### 3.4.8. Experimental research facilities for laser fusion

A typical experimental facility for laser fusion is shown in Fig. 3.32. Laser light, which is emitted from the oscillator shown on the right and lower part of the figure, assumes a Gaussian shape in tune with the pulse width of 100 ps, after it passes through the pulse-shaping stage. Three half-mirrors M1, M2 and M3 separate the light into four beams. After the beam intensity is increased by passing through amplifiers I, II, III, IV and V, the beam is focused by the lens on the target surface. On the way from the oscillator to the target,

**Fig. 3.32.** Experimental facility for laser fusion.

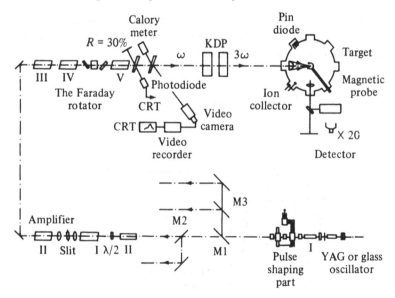

several slits and a Faraday rotator are located. The slits homogenise the beam intensities, while the Faraday rotator protects the amplifier from damage due to reflected light. The KDP induces the third harmonic of Nd laser light. Since the beam intensity is limited to $1 \text{ J/cm}^2$ or so, the light energy per beam is several hundred J. To increase the beam energy, the number of beams must be increased. At present, there is an experimental facility with 20 beams in operation.

The target, a glass microballoon, is shown in Fig. 3.33. The glass shell must be a true sphere and its thickness must be homogeneous. The cryogenic target includes the D–T fuel in the solid state. As shown in the figure, the central part of the target is void, in order to increase the compression of the fuel by implosion. For spherically-symmetric implosion of the target, the laser light must irradiate the target in a spherically-symmetric way. But 20 beams cannot irradiate the target surface with true spherical symmetry. A low-$Z$ material such as polyethylene coated on the glass microballoon absorbs laser light well and homogenises the nonuniform temperature (and pressure) on the target surface because of its high thermal conductivity.

Using this kind of target, however, the greater part of the laser energy absorbed in the target escapes from the target via the materials blown off it. Only a small target gain can thus be expected.

The cannonball target shown in Fig. 3.34 protects the target surface from being blown off. The surface of the cannonball target consists of two shells of a high-$Z$ metal, with an unfilled gap between the shells. The outer shell has two holes through which the laser light can pass. Light is reflected by both shells many times, and the

**Fig. 3.33.** Target structure.

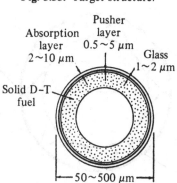

shells radiate soft X-rays which fill the vacant gap. The high-$Z$ metal is then vaporised and the vapour fills the gap at high pressure, which induces the implosion of the fuel. The outer shell plays the role of a tamper, preventing the target from being blown off the surface material.

Currently the efficiency of such a laser is too low, and hence the total laser energy is too small for laser fusion. The efficiency of laser light is one of the major problems in fusion research.

## 3.5. Electron beams

### 3.5.1. Relativistic electron beams

In contrast to the low efficiency of the laser, the electron beam has high energy conversion efficiency from electrical energy to electron kinetic energy. Due to the recent rapid development of the pulsed-power technique, a field of brilliant high technology, the electron beam is now able to serve as the energy driver for inertial-confinement fusion. About 1 cm from the metallic cathode, a mesh anode is located in a vacuum ($10^{-4}$ Torr), and a high voltage (several MV) is applied to the anode. Electrons, extracted from the cathode, are accelerated toward the anode and pass through the mesh anode, forming the electron beam (Fig. 3.35). The energy conversion rate from the input electrical energy to the kinetic energy of the electron beam exceeds 60 %. To ensure that the beam width is of the order of 10 ns, the pulse width of the high voltage applied to the anode must also be of order of 10 ns.

Fig. 3.34. A canonball target.

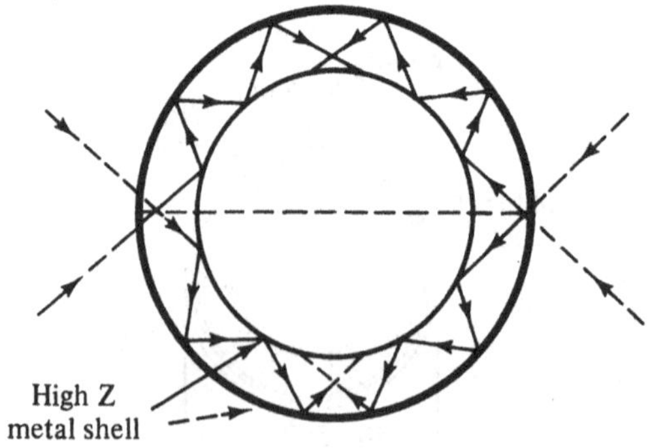

High Z
metal shell

The kinetic energy of an electron is several MeV which is equal to the corresponding voltage between the electrodes. If the rest mass of the electron is denoted by $m_e$, the velocity of light by $c$, and the kinetic energy of an electron by $E_e$, then $E_e$, given by

$$E_e = \tfrac{1}{2}m_e c^2, \tag{3.131}$$

is about 300 keV. In other words, the electron accelerated by the voltage of 300 kV reaches the velocity of light. Generally, the electron beam for inertial fusion is accelerated by a voltage of several MV. The velocity of the electron remains at that of light, but its mass increases according to eq. (1.6). Thus the electron beam is called the relativistic electron beam (REB).

### 3.5.2. The Marx generator

The high-voltage power supply system required to extract the REB is called the Marx generator. In Fig. 3.36, the A switches are first on and the B switches are off. The capacitor banks are then charged in parallel by the high-voltage source. Next, the A switches are turned off (in practice, the A switches are wires with a high impedance and are not required to be off for the pulsed current) and the B switches are turned on. The capacitor banks are then reconnected in series. If the charging voltage is 100 kV for the 100-capacitor

**Fig. 3.35.** Diode for the electron beam.

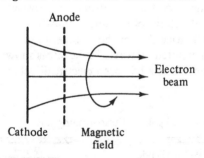

**Fig. 3.36.** Marx generator.

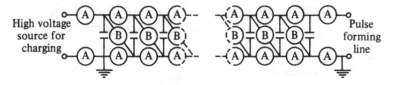

banks of capacitance 5.4 $\mu$F, the energy stored in the Marx generator will be 2.7 MJ and the output voltage 10 MV, after series connection. The high-power pulses are then sent to the pulse-forming line.

If 12 modules of such a Marx generator are constructed in this way, the total stored energy reaches 32 MJ. In order that the many modules of the Marx generator operate simultaneously, the B switches in Fig. 3.36 must close simultaneously. The jitter of these B switches must be controlled so as to be less than 2 ns, since the pulse width of the power system is of the order of tens of ns. If the first B switch closes, the voltage pulse closes the other B switches successively. Thus for the synchronisation of the multi-modules of the Marx generator, the first B switch of each module is operated by a synchronised triggered signal.

### 3.5.3. *Intermediate storage capacitors and the pulse-forming line*

Two coaxial cylindrical metal layers filled with pure water make up what is called an intermediate storage capacitor. The capacitor bank has a high capacity because water has a high dielectric constant. Since the breakout voltage of the water insulator depends on the temporal width of the high-voltage pulse, the pulsed power from the Marx generator is first stored in the intermediate storage capacitor, and then sent to the pulse-forming line as a shorter power pulse.

The pulse-forming line shown in Fig. 3.37 consists of coaxial cylindrical layers. A single-phase coaxial line and a Blumlein line are generally used in pulse forming, and the pure water with its high dielectric constant serves as the insulator between the cylindrical layers. The output voltage of the single-phase coaxial line is half the input voltage and twice that of the Blumlein line.

The transition of potential waves in the line is shown schematically in Fig. 3.38. After the line has been electrically charged, one end is

**Fig. 3.37.** Pulse-forming line.

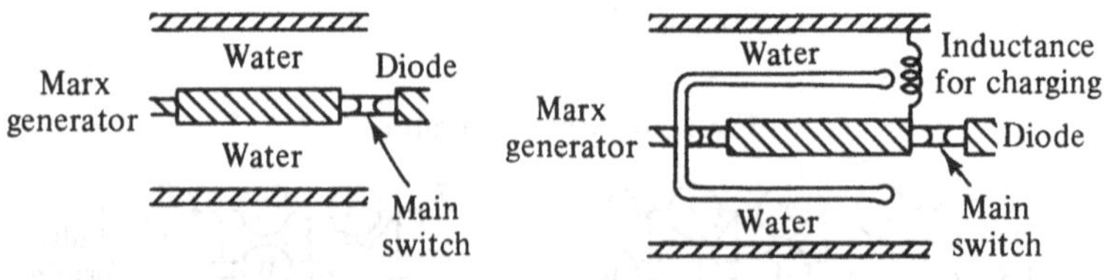

(a) Single phase          (b) Blumline line
coaxial line

closed. Two potential waves then propagate in opposite directions
to each other. The positive wave is reflected as positive at the open
end, and the negative wave is reflected as negative at the closed end.
In the figure, (a) shows the wave propagating to the right, (b) the
wave propagating to the left, and (c) the combined potential. The
propagation velocity is denoted by $c$ and the length of the line by
$l$. At the instant $t = 0$ shown in Fig. 3.39(a), electrode B is earthed,
electrode A is charged positively against B, and electrode C is charged
positively against B. The initial electric fields between the electrodes
are shown in Fig. 3.39(a). If the gap at the left end between A and
B is closed, the potential is inverted when $t = 6l/3c$, and the electric
fields become as in Fig. 3.39(b). The potential difference between A
and C is just twice that between B and C. At that moment, the main
switch in Fig. 3.37(b) is closed. Then the high-voltage pulse transfers
to the diode.

### 3.5.4. The impedance conversion line and the plasma-erosion opening switch

It is possible to supply a high voltage of 10 MV to the diode directly
from the pulse-forming line. However, there is also a convenient
method of using the impedance conversion line to increase the output
voltage. Suppose the output voltage of the pulse from the

**Fig. 3.38.** Propagation waves at the electrode of the pulse-
forming line.

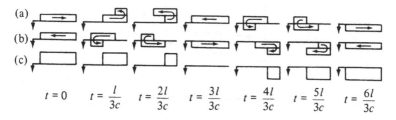

$$t = 0 \qquad t = \frac{l}{3c} \qquad t = \frac{2l}{3c} \qquad t = \frac{3l}{3c} \qquad t = \frac{4l}{3c} \qquad t = \frac{5l}{3c} \qquad t = \frac{6l}{3c}$$

**Fig. 3.39.** Electric fields in the Blumlein before and after the
shorting of one end.

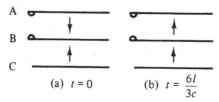

(a) $t = 0$ \qquad (b) $t = \dfrac{6l}{3c}$

pulse-forming line is 3.3 MV. The pulsed power is sent to the impedance conversion line shown in Fig. 3.40. When the impedance of the line at the entrance section is 1 Ω (which must be matched to the impedance of the pulse-forming line), while the impedance at the exit section is 9 Ω, the output voltage from the line is raised to 10 MV (three times the input voltage of 3.3 MV).

Plasma is projected from the plasma gun into the transmission line between the pulse-forming line and the diode (Fig. 3.41). There the plasma forms a short circuit in front of the diode, and the pulsed-power energy is converted to magnetic energy in the transmission line. Plasma erosion then opens the circuit after a short period, with the help of the strong magnetic pressure, and the magnetic energy stored in the transmission line is converted back into the pulsed power of the diode. By means of such a plasma-erosion opening switch, the voltage of the pulsed power is raised three times and the pulse width is shortened to one half.

### 3.5.5. Diode

To focus the beam to a point, the electrodes take the form of concentric spherical layers. However, for simplicity, the case of the

**Fig. 3.40.** The impedance conversion line.

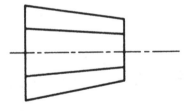

**Fig. 3.41.** Plasma-erosion opening switch.

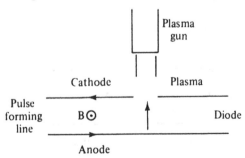

plane electrodes will be examined here. Let us observe the electron current extracted from a cathode of unit area (Fig. 3.42).

The phenomenon is assumed to be steady and depends only on $x$. The potential difference between the anode and the cathode is denoted by $V$ and the potential of the cathode and the anode by $\Phi_C = 0$ and $\Phi_A = V$ respectively. On the cathode surface, from which the electrons are extracted, the relations $n_e = \infty$ and $d\Phi/dx = 0$ are assumed to be satisfied. In the gap between the anode and the cathode, only electrons extracted from the cathode are to be found. The number density of electrons at $x$ is written as $n_e(x)$, and the potential there by $\Phi(x)$. The Laplace equation for $\Phi$ is

$$\varepsilon \frac{d^2\Phi}{dx^2} = en_e, \qquad (3.132)$$

where $\varepsilon$ is the dielectric constant of a vacuum and $-e$ is the charge of the electron. On the other hand, the flight velocity of the electron in the gap is denoted by $v(x)$, and the continuity equation for the electron (for the current) is

$$en_e v = \text{constant} \equiv j_e, \qquad (3.133)$$

where $j_e$ is the current density of the electron. At $x = 0$ we have $v(0) = 0$. The energy equation for the electron is

$$\tfrac{1}{2}m_e v^2 = e\Phi.$$

Using eq. (3.133), eq. (3.132) is rewritten as

$$\frac{d^2\Phi}{dx^2} = \frac{j_e}{\varepsilon}\left(\frac{m_e}{2e\Phi}\right)^{\frac{1}{2}}. \qquad (3.134)$$

**Fig. 3.42.** Diode and potential.

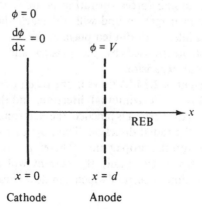

$\phi = 0$

$\dfrac{d\phi}{dx} = 0$

$\phi = V$

REB

$x$

$x = 0$      $x = d$

Cathode      Anode

Taking $\Phi = 0$ and $d\Phi/dx = 0$ at $x = 0$ into account, we integrate eq. (3.134) twice to give

$$\Phi = \left[\frac{9j}{4\varepsilon}\left(\frac{m_e}{2e}\right)^{\frac{1}{2}}\right]^{\frac{2}{3}}x^{\frac{4}{3}}.$$

If the relation $\Phi = V$ at $x = d$ is substituted into the above equation, we have

$$j_e = \frac{4\varepsilon V^{\frac{3}{2}}}{9d^2}\left(\frac{2e}{m_e}\right)^{\frac{1}{2}}. \tag{3.135}$$

As eq. (3.135) indicates, the electron current $j_e$ is proportional to $V^{\frac{3}{2}}$ and to $d^{-2}$. The electron current $j_e$ given by eq. (3.135) is called the Child–Langmuir current. When the stored energy in the Marx generator is 2.7 MJ and the potential difference between the electrodes is 10 MV, an electron beam energy of 1 MJ during a pulse length of $t = 6l/3c = 100$ ns corresponds to a beam current $I_b = 2.7 \times 10^6$ A. Equation (3.135) indicates that $j_e = 7.4 \times 10^8$ A/m$^2$ when $V = 10$ MV and $d = 1$ cm. Thus the area of the cathode is $S = I_b/j_e = 3.6 \times 10^{-3}$ m$^2 = 36$ cm$^2$.

### 3.5.6. *Propagation of relativistic electron beams*

The REB which is projected by the diode irradiates the target containing the D–T fuel. The reactor, in which the fusion reactions are to occur, has a spherical wall whose radius is about 5 m. The beam is extracted from the diode outside the reactor and propagated into the reactor cavity to the target at the centre of the reactor. In other words, the electron beam must propagate a distance of 5 m. For a strong current of 2.7 MA, the number density of the beam electron is large. Since the electrons are scattered away from each other by the Coulomb forces operating between them, the beam expands during propagation and will not reach the target. If the reactor cavity is filled with diluted plasma, the charge of the beam electron will be neutralised by the plasma and the beam will propagate without expansion.

When the current of 2.7 MA flows in the reactor cavity, it induces a magnetic field $B_\theta$ in the azimuthal direction, and this field pinches the current. When the current is pinched, the electrons have a velocity component $v_r$ in the radial direction. The Lorentz force $v_r B_\theta$ causes the electrons to stop the propagation. Therefore, when the electron current is too strong, eventually the current will not be able to advance. The maximum current which can flow is called the Alfvén limit.

The REB, extracted from the diode with the potential difference of $V$, circulates by action of the Lorentz force due to the self-induced magnetic field. If the radius of circulation is less than the radius of the beam, the beam cannot propagate. Thus the Alfvén limit $I_A$ is given by

$$I_A = 1.7 \times 10^4 V^{\frac{1}{3}} \text{ A} \qquad (V \text{ in MV}). \qquad (3.136)$$

In practice, a beam current exceeding the Alfvén limit can in fact propagate. The reason is as follows (Fig. 3.43). The beam current induces a magnetic field $B_\theta$ around the current. The increase in $B_\theta$ in time induces an electric field $E_z$ in the propagation direction. A back current, which cancels the beam current, flows due to $E_z$. The effective beam current is decreased by the back current. For an electron, however, its charge is large in comparison with its mass. The scattering by the Coulomb force is strong when the beam charge is not neutralised. When the charge is neutralised, the strong current induces the pinch motion. Thus one of the disadvantages of the REB is the difficulty of propagation. The beam path is covered by a cylindrical metal layer, in which the back current flows easily, and the coil wound around the metal layer induces the magnetic field $B_z$ in the propagation direction, which guides the electron beam to the target. In a practical fusion reactor, however, this kind of beam cover and coil cannot be located in the cavity.

### 3.5.7. *Coulomb scattering of the beam at the target*
When the REB impinges on the target, the target surface absorbs the beam energy to be ionised. The REB interacts strongly with electrons in the target, and is slowed to a stop, transferring energy to electrons in the target. The REB passes through the Coulomb fields formed by electrons and ions in the target, and its path is bent, until eventually it is scattered in many directions. When the beam

**Fig. 3.43.** REB propagation.

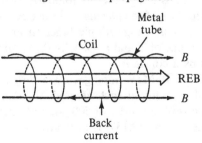

velocity is large, the interaction is weak because the duration of interaction is short due to the high velocity. The stopping power of the target with respect to the electron beam is inversely proportional to the square of the beam velocity (inversely proportional to the beam energy).

If the kinetic energy of an electron in the beam is denoted by $E_b$ ($E_b = \frac{1}{2} m_e v^2$), then the stopping power $dE_b/dx$ is expressed by

$$\frac{dE_b}{dx} = -\frac{e^2}{4\pi \varepsilon v^2} \sum_* \log \Lambda \cdot n^* \frac{q^{*2}}{m^*}$$
$$\times \left[ \Phi(b^* v) - \left(1 + \frac{m^*}{m_e}\right) \frac{2b^* v}{\pi^{\frac{1}{2}}} \exp(-b^{*2} v^2) \right], \quad (3.137)$$

where

$$b = \frac{m}{2T}, \qquad \Phi(x) = \frac{2}{\pi^{\frac{1}{2}}} \int_0^x \exp(-\xi^2)\, d\xi, \qquad (3.138)$$

in which $b$ is the inverse of the square of the thermal velocity and $\Phi$ is the error function. In eq. (3.137), the quantities marked with an asterisk indicate electrons or ions in the target plasma, while $\sum_*$ indicates the sum of the stopping powers of an electron and an ion. Since the denominator on the right-hand side of eq. (3.137) includes $m^*$, the electron, whose mass is small, contributes significantly to the stopping power. And since electron mobility is much higher than that of an ion, electrons in the wide region are affected by the impinging electron beam. In reaction, the electrons in the plasma stop the beam. In eq. (3.137), $\log \Lambda$ is called the Coulomb logarithm and is given by

$$\log \Lambda = \log \frac{3\varepsilon}{Ze^3} \left(\frac{4\pi k^3 T_e^3}{n}\right)^{\frac{1}{2}}, \qquad (3.139)$$

where $Z$ is the atomic number of the ion in the plasma and $k$ is the Boltzmann constant. Although the Coulomb logarithm is a function of the number density $n$ of the plasma and the electron temperature $T_e$ (see Table 2.1), it is a logarithmic function and has values of 10–20 for a wide range of $T_i$ and $n$. Usually the Coulomb logarithm is approximated by a constant.

When an REB of 10 MeV impinges on a target surface which consists of a high-$Z$ material such as gold, the stopping range of the electron beam is of the order of 500 $\mu$m, although the stopping range as calculated by eq. (3.137) is of the order of 5 cm.

### 3.5.8. Abnormal interaction of REB with the target

Experimentally it has been observed that when an REB of 1 MeV impinges on metal film, the REB is stopped within a thickness of 1 μm, which is much less than the stopping range calculated by eq. (3.137). This indicates that the REB undergoes a strong interaction with the plasma in addition to the Coulomb interaction.

The two-stream instability is a well-known micro-instability. The REB in the target has a high velocity relative to electrons in the target plasma. In Fig. 3.44, two groups of electrons move with relative velocity $v$ against each other (the frame of reference moves with the velocity $v/2$ relative to the laboratory frame). Suppose that the electrons have density fluctuation $n'_e$ and hence that an electric field $E$ appears in the x-direction. An electron will be accelerated or decelerated by this electric field. The point of maximum velocity of the electron is $P$, and the point of minimum velocity is $Q$. Accordingly, the electron density at $P$ decreases and the density at $Q$ increases. Thus the instability grows.

The ratio of the number density of electrons moving to the right to those moving to the left is denoted by $\alpha$. When $\alpha \ll 1$ (the number density of the REB is much smaller than that of the target plasma), the growth rate $\gamma$ of the two-stream instability is

$$\gamma = \frac{3^{\frac{1}{2}}}{2}\left(\frac{\alpha}{2}\right)^{\frac{1}{3}}\omega_{\mathrm{p}}, \qquad (3.140)$$

where $\omega_{\mathrm{p}}$ is the plasma frequency. The plasma frequency is directly related to $n_e$ and has a value of $10^{12}$ rad/s for the target plasma. Accordingly, the growth rate $\gamma$ of the two-stream instability has a large value in the target, and is able to grow during the REB's pulse width of the order of $10^{-8}$ s. Equation (3.140) can be applied for the case of a REB with a low temperature. When the temperature of the REB is high, the growth rate $\gamma$ of the two-stream instability

**Fig. 3.44.** Two-stream instability.

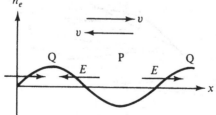

is given by

$$\gamma = \left(\frac{\pi}{8}\right)^{\frac{1}{2}} \alpha \frac{(v - v_T)^2}{v_T^2} \omega_p e^{-\frac{1}{2}}, \qquad (3.141)$$

where $v$ is the velocity of the REB and $v_T$ is the thermal velocity, which is related by the temperature of the REB by

$$v_T = \left(\frac{2kT_b}{m_e}\right)^{\frac{1}{2}}. \qquad (3.142)$$

When $v_T = v$, no two-stream instability grows.

The two-stream instability is based on the electrostatic wave (only the electric fields appear in the direction of propagation of the wave). In a similar situation, the Weibel instability (electromagnetic mode) is manifested. When two groups of electrons move against each other in the $x$-direction with the velocity $v$, an electromagnetic wave (with magnetic field in the $z$-direction and electric field in the $x$-direction) whose wave number is in the $y$-direction, is taken to be induced. The path of the electron beam is bent by the induced magnetic field, as shown in Fig. 3.45. Thus the electron beam forms local currents which enhance the induced magnetic field. The growth rate $\gamma$ of the Weibel instability is given by

$$\gamma = \left(\frac{8}{27\pi}\right)^{\frac{1}{2}} \frac{\omega_p}{c} \left[ (1 - \alpha) \frac{v_{PT}^2 + v^2}{v_{PT}^2} + \alpha \frac{v_T^2 + v^2}{v_T^2} - 1 \right]^{\frac{3}{2}}$$

$$\times \left[ (1 - \alpha) \frac{v_{PT}^2 + v^2}{v_{PT}^2} + \alpha \frac{v_T^2 + v^2}{v_T^3} \right]^{-1}, \qquad (3.143)$$

where $c$ is the velocity of light in a vacuum and $v_{PT}$ is the thermal velocity of electrons in the target plasma.

**Fig. 3.45.** Weibel instability.

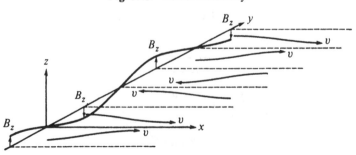

Because the growth rates of the two-stream instability and the Weibel instability are large, the electrostatic wave and the electromagnetic wave grow in the target soon after REB irradiation, and the kinetic energy of the REB is transferred to the energy of the growing fluctuations. The energy of the fluctuations is eventually converted to thermal energy of the electrons. Thus the REB is stopped strongly in the target, and the target surface is heated rapidly.

Currently the mechanism of the stopping of the REB is not clear. In addition to the micro-instabilities described above, the magnetic field seems to play an important role in stopping the REB (Fig. 3.46). REB propagation in the reactor cavity induces a strong magnetic field $B_\theta$ around the beam. This magnetic field soon impinges on the target. The beam electrons are trapped by this magnetic field in the target surface, and rotate around the magnetic field without penetrating the deep layer. As a result, the REB deposits its energy on the surface of the target.

Although the REB has the advantage that the rate of energy conversion is remarkably high, it has two main disadvantages. One is the difficulty of propagation. At present, no solution has been found for this problem. The second is the abnormal interaction of the REB with the target. Like the absorption of laser light, the abnormal phenomena are very complex and not clear. However, energy absorption on the thin surface of the target induces blowing off of the target material, thus decreasing target gain.

## 3.6. The light-ion beam

### 3.6.1. Characteristics of the light-ion beam

Using a light-ion beam (LIB) as the energy driver for inertial-confinement fusion has many advantages in comparison with

Fig. 3.46. The magnetic field near the target surface induced by the REB.

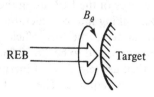

the REB. The LIB consists of ions with low atomic number, such as protons or Li. Like the REB, the LIB can be accelerated by a diode with a voltage of 1–10 MeV. The efficiency of energy conversion from electrical to kinetic energy of the beam is not too low, though of course it is lower than that of the REB. Since the ion is more than several thousand times heavier than the electron, the effect of the charge is much smaller on the ion than on the electron. The velocity of the LIB does not reach that of light when it is accelerated by a voltage of 10 MV. Thus an electron, which has high mobility, can follow the motion of the ion and cancel out the charge effect of the ion. Scatterings by their own Coulomb fields and pinch motion during the propagation are therefore much weaker for the LIB in comparison with the REB. Since the propagation velocity of the LIB is less than the velocity of light, its propagation velocity changes according to the voltage of the diode. Beam bunching will be achieved if we can control the voltage between the electrodes of the diode.

The LIB interacts with the target via the Coulomb force only. The preferable stopping range (100 $\mu$m) of the LIB in the target can be achieved by using the appropriate diode voltage (4 MV for a proton and 30 MV for Li) and it may be expected to have a high target gain. Since the ion mass is much larger than the electron mass, the ion momentum is larger than the electron momentum if the particle energies are comparable. There is a possibility that the LIB's momentum assists in imploding the fuel into the target.

The reaction probability $\langle \sigma v \rangle$ has a maximum at $T = 100$ keV for the D–T reaction, which is shown by Fig. 1.4. When a D ion beam impinges on the T fuel with a velocity corresponding to 100 keV, direct fusion (not the thermal motion but directly by kinetic energy) will release the energy.

With respect to the advantages of the LIB, research on inertial-confinement fusion is concentrating on the LIB, along with the REB.

### 3.6.2. Generation of the LIB

The efficiency of the energy conversion from the electrical input energy to the kinetic energy of the LIB, using the diode, is expected to be more than 50 %. Taking into consideration the loss of energy in the power-supply system, the net efficiency of the energy conversion of the LIB is expected to be 25–30 %. In order to obtain fusion energy for practical use (e.g. 3 GJ of thermal output), a beam energy of 8 MJ per shot is required. The input electric energy for this order of beam energy is required. The input electric energy for this order of beam energy is 32 MJ. The power-supply system consists of a Marx generator and a pulse forming line, the same as for the REB.

An example of the power-supply system is illustrated in Fig. 3.47; the cross-section of the system is circular. The three types of power-supply systems to be considered here are summarised in Table 3.2. For a power-supply system of type 1, which supplies an output voltage of 10 MV, the following parameters are chosen, in consequence of the development of the plasma-erosion opening switch:

*Marx generator:* Capacitance of condensers $C = 135\ \mu F$ (parallel connection of smaller capacitor banks is acceptable) in oil; charging voltage $V_c = 200\ kV$; number of stages $N_c = 12$; output voltage $V_m = 2.4\ MV$; total storage energy $E_m = 2.7\ MJ$.

*Pulse-forming line:* Single-phase coaxial line using pure water as insulator; pulse width $t_p = 60\ ns$; output voltage $V_p = 1.2\ MV$.

*Impedance conversion line:* Entrance impedance $Z_{en} = 0.3\ \Omega$; exit impedance $Z_{ex} = 2.7\ \Omega$; output voltage $V_i = 3.6\ MV$.

*Plasma-erosion opening switch:* Pulse width $t_{pu} = 30\ ns$; output voltage $V_{pu} = 10\ MV$.

Table 3.2 *Power-supply systems*

| Type | Stored energy | Output voltage | Pulse width | Number of modules |
|------|--------------|----------------|-------------|-------------------|
| 1 | 2.7 MJ | 10 MV | 30 ns | 6 |
| 2 | 1.7 MJ | 5 MV | 30 ns | 6 |
| 3 | 2.6 MJ | −1 MV | 200 ns | 1 |

**Fig. 3.47.** A power-supply system, including a Marx generator, an intermediate storage capacitor, a gas switch, a pulse-forming line, a magnetic switch, and an impedance conversion line.

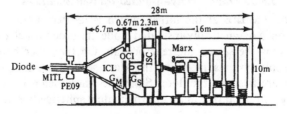

If the gas switches in these systems are triggered by laser light, and water switches are replaced by magnetic switches, the jitter of the switches becomes less than 2 ns and the power-supply systems can be operated with synchronisation.

One power-supply system may be divided into two groups. From one group, a pulsed electric power of pulse width 30 ns, output voltage 10 MV, and stored energy of 2.7 MJ can be obtained. From the other group comes a pulsed electric power of pulse width 30 ns, output voltage 5 MV and stored energy 1.7 MJ. As explained in the next section, two kinds of ion beam, rotating and non-rotating ones, may be combined to form one beam. This gives rise to the two kinds of power-supply system described above. The efficiency of power transmission from the Marx generator to the intermediate-storage capacitor bank, and from the intermediate-storage capacitor bank to the pulse-forming line, are estimated to be 90 %, respectively. If the efficiency of the impedance-conversion line is 80 %, the total rate of energy transmission inside the power-supply system is 64.8 %.

If the energy conversion efficiency at the diode from electrical energy to the LIB's kinetic energy is 50 %, the total driver efficiency is $\eta_d = 32.4$ %. In what follows, we underestimate the driver efficiency at $\eta_d = 25$ %.

Power-supply systems of type 1 and type 2 supply electric power to the diodes in order to extract LIB. Power-supply systems of type 3 are used to apply the biasing voltage to the target, as will be described later.

For inertial-confinement fusion, construction of the laser facility to extract 8 MJ is almost impossible, or at any rate very expensive. On the other hand, the capacitor banks that constitute the main part of the LIB fusion machine are available at a reasonable price. The 4 MJ beam machine, PBFA II, in the Sandia National Laboratories designed for LIB fusion has been constructed for US$20 million. The established technology and the reasonable cost (1/200 of that of the Tokomak) of the TFTR at Princeton University, promises a bright future for research into LIB fusion.

### 3.6.3. Magnetically-insulated transmission lines

To project the LIB at the target, which is located at the reactor centre, from a magnetically-insulated diode, the diode which extracts the LIB must be set on the reactor wall. On the other hand, the power supply systems, including the Marx generators, are so heavy that they are fixed on the floor. Since the whole beam is expected to impinge on the target in an approximately spherically-symmetric way, the diodes and beam ports are distributed at spherically-symmetric positions on the cavity wall. Thus the lengths of the transmission line between the power supply systems and the diodes are not equal. The firing times of the Marx generators must be controlled so that the synchronised beams impinge on the target. When the transmitted current $I$, flowing in the coaxial cylindrical line shown in Fig. 3.48, exceeds a critical value, electrons which are extracted from the inner conductor (cathode) by the high voltage cannot reach the outer conductor (anode), because they are bent by the magnetic field which is produced by the transmitted current itself. Thus the voltage needed to break the insulation of the transmission line is increased by the self-induced magnetic field. This type of line is called a magnetically-insulated transmission line.

To extract the LIB from the diode on the reactor wall, the polarity of the electrode for the LIB must be inverted from that of the REB. But by inverting only the polarity of the electrodes, the electron current also flows in the diode and the ion current cannot be extracted. It is therefore necessary that the diode has a mechanism to suppress the electron current. Figures 3.49 and 3.50 show a

**Fig. 3.48.** A magnetically-insulated transmission line.

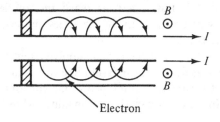

**Fig. 3.49.** Schematic diagram of a magnetically-insulated diode.

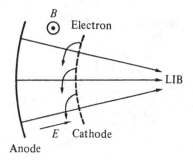

magnetically-insulated diode in which the externally applied magnetic field suppresses the electron current and assists in extracting the ions from the anode. The electrons in the diode undergo drift motion in the direction of $\mathbf{E} \times \mathbf{B}$ and cannot reach the anode. On the other hand, the Larmor radii of ions in the diode are much larger than those of electrons. The ions are accelerated toward the cathode, scarcely affected by the magnetic field, and pass through the cathode to form an ion beam. Near the cathode, electrons neutralise the charge of the beam ions.

The velocity of the proton beam accelerated by the voltage of 10 MV is $4.4 \times 10^7$ m s$^{-1}$, which is about $\frac{1}{10}$ of the velocity of light. If we assume that there are only ions between electrodes of the diode, the current density $j_p$ of the proton beam extracted from the diode is the Child–Langmuir value given by eq. (3.135), if we replace the electron mass $m_e$ by the proton mass $m_p$. In a real diode, there are electrons near the cathode. Because of the existence of these electrons, the effective gap distance between the anode and cathode of the diode reduces from the geometrical gap distance. Hence the proton current density $j_p$ increases from the Child–Langmuir value. This enhancement factor of the current density depends on the intensity of the magnetic field, and the factor increases with a weaker magnetic field. Of course, the intensity of the magnetic field must exceed the critical value; if it is less, electrons reach the anode. Usually a magnetic field with an intensity of 1.5 times the critical value is applied to the LIB diode.

The pulsed power is sent from power supply system 1 or 2 through the magnetically-insulated transmission line to the diode set on the wall of the reactor cavity. A combined diode consists of two parts (see Fig. 3.51). In the outer part there is an anode whose outer radius

**Fig. 3.50.** An example of a magnetically-insulated diode.

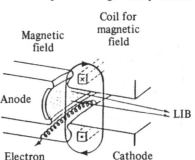

is 34 cm and inner radius 30 cm. (The impedance of the diode must match the impedance 2.7 Ω at the exit section of the impedance-conversion line.) Over a range of 6.4 cm, including the gap distance of 6.3 mm between the anode and the cathode, a radial magnetic field of average intensity 5.1 T is applied. The beam ions from the anode must be focused to a small radius by the geometrical focusing method of the diode or by an externally-applied magnetic field in the azimuthal direction. Once the beam is focused, the self-induced magnetic field confines the beam to this small radius. From the point of view of beam interaction with the target, it is desirable that the particle energy of the beam should increase with time, because the stopping powers of the target material for the beam ions increase as their temperatures increase, as shown in section 3.6.5. (It is preferable that the stopping range of the beam remains constant in the target with respect to time, independent of the temperature of the target material.) The diode voltage (i.e. the particle energy of the beam) during the beam extraction depends on the intensity of the magnetic field which insulates the electron current in the diode. Thus our purpose will be achieved when the coil which induces the insulation magnetic field is connected in series to the diode line, because the intensity of that field increases with time, accompanied by the increasing diode current. By using this kind of diode, there is no insulation magnetic field between the electrodes at the initial stage of the diode operation. Therefore the electrons from the cathode hit the anode surface in this initial stage without being hindered by the magnetic field. This initial electron irradiation of the anode surface will produce the anode plasma, which supplies the subsequent

**Fig. 3.51.** Combination of two diodes. A = Anodes, C = cathode and *B* = magnetic fields.

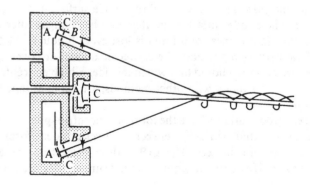

ion current with uniform density from the large anode surface. When pulsed power of 10 MV is supplied to the diode with an intensity of 5.1 T of the magnetic field to insulate the electron current, a proton beam of current density of 5 kA cm$^{-2}$ is extracted, rotating around the propagation axis by the action of the Lorentz force of the radial magnetic field in the diode. The propagation and rotation energies of the proton beam become 4 MeV and 3 MeV, respectively (with thermal energy of 1 MeV). The total current of the rotating beam along the propagation direction is 1.9 MA. From an ordinary non-rotating proton beam of the total current of 5 MA (current density 5 kA cm$^{-2}$) is extracted, with the particle energy of 4 MeV, the most desirable propagation particle energy for protons from the point of view of the beam–target interaction. Since the rotating beam forms a hollow ring, another non-rotating beam must fill the hollow part of the ring beam.

Because a current density of 8 kA cm$^{-2}$ has already been extracted from the anode surface, it can reasonably be expected that 5 kA cm$^{-2}$ may be extracted. In many diodes, the anode surface of the diode is covered by the polyethylene, which includes hydrogen in high concentration. When the high voltage is applied between the electrodes, the anode plasma is formed by edge-surface discharge. The ions in the anode plasma are accelerated to form an ion beam. In the case where the formation of the anode plasma is late, the proton beam is delayed by the high voltage pulse, and the beam power becomes small. Many knock pins, which have sharp edges to form local strong electric fields, are fixed on the anode surface, or the anode surface accepts the short electron bombardment or laser irradiation, assisting the edge-surface discharge to take place easily without time delay.

The anode surface evaporates to decrease the thickness 0.1 $\mu$m per shot. If the operation frequency is 1 Hz, there can be $3 \times 10^7$ shots in one year. The thickness of the anode surface decreases 3 m per year. The anode surface must therefore be changed after every 1000th shot (a thickness of 0.1 mm is lost every 3 hours). Methods of puffing hydrogen gas onto the cryogenic anode, or covering the anode surface with a liquid film which includes a high concentration of the hydrogen, along with other methods, are being examined for the diode.

The electrode surfaces have the form of concentric spheres, so that the equi-potential surface between the electrodes forms the concentric sphere surfaces. The LIB is then focused on the centre of the sphere. However, the equi-potential surfaces deviate from the spherical near the edge of the diode, the electric fields have ripples near the knock pins, and the magnetic fields, externally-applied and self-induced, bend ion paths. Because of these factors, the ion beam cannot be focused to a point, but has some divergence angle of the order of 1 degree. If we denote the spot size of the beam at the focused point by $r_0$, the divergence angle of the beam by $\theta$, and the focusing length by $f$, we have (see Fig. 3.52)

$$f \cdot \theta = 2r_0. \tag{3.144}$$

If we assume that $r_0 = 4$ mm and $\theta = 1$ deg $= 0.017$ rad, then we have $f = 47$ cm. When the radius $r_A$ of the anode is 6 cm, the maximum incident angle $\varphi$ of the beam is $\varphi = 6/47$ rad $= 0.13$ rad $= 11$ deg.

### 3.6.4. Propagation of the LIB

The LIB has the advantage that it is easily accelerated by a diode. On the other hand, it has the disadvantage that it has a divergence angle and focusing length of about 50 cm.

As will be described in chapter 4, since the radius of the reactor is of the order of 5 m at least, it is difficult to focus the LIB projected from the diode on the reactor wall onto the target, whose radius is less than 1 cm. A method which uses a plasma channel has been proposed to propagate the LIB in a reactor cavity which is filled with gas.

The inside of the reactor is filled with an inert gas, such as neon or argon, at a pressure of 10 Torr. Through electrodes on the reactor wall, a discharge occurs in the gas, forming a plasma channel. It is difficult to discharge the gas directly into the reactor, because its

Fig. 3.52. Focusing an LIB from the diode.

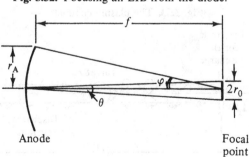

Anode

Focal point

diameter is 10 m. If laser light passes through the gas, the outer
bound electrons of the gas molecules are excited, even though the
laser light is not strong enough to ionise the gas. As the electrons
of the gas molecules are excited, the voltage needed to discharge the
gas decreases significantly.

The discharge current induces a magnetic field $B_\theta$ in the azimuthal
direction around the current. This magnetic field is expected to
confine the LIB in the plasma channel. In Fig. 3.53, the focal radius
of the LIB from the diode is denoted by $r_0$, the maximum incident
angle by $\varphi$, the largest radius of the LIB propagation in the channel
by $r_c$, and the beam energy by $V$. The discharge current $I_c$, by which
the LIB is confined in the plasma channel, is given by

$$I_c = 1.65 \times 10^{-10} r_0 V \varphi n_b^{1/2} \text{ A} \qquad (\varphi \text{ in rad; } V \text{ in MeV}). \qquad (3.145)$$

If we assume that $V = 10$ MeV, $\varphi = 0.13$ rad, $r_0 = 4 \times 10^{-3}$ mm, and
the beam number density $n_b = 1.29 \times 10^{20}$ m$^{-3}$, then $I_c = 67$ kA. The
maximum magnetic field $B_\theta$ induced by this discharge current at
$r_0 = 4 \times 10^{-3}$ mm is 59 T.

The neighbouring plasma channels are connected to each other
in order to let the discharge current flow easily, as shown in Fig.
3.54. The inductance $L$ of the circuit is about 5 $\mu$H. If we suppose
that the electron temperature of the plasma channel is 5 eV, the
radius of the channel 6 mm, and the length of the channel 10 m,
then the electrical resistance of the circuit is about $R = 15 \ \Omega$. The
circuit consists of the series connection of $L$, $R$ and $C$ (Fig. 3.55).
The current $J$ which flows in the channel is governed by

$$L \frac{d^2 J}{dt^2} + R \frac{dJ}{dt} + \frac{J}{C} = 0, \qquad (3.146)$$

**Fig. 3.53.** The plasma channel.

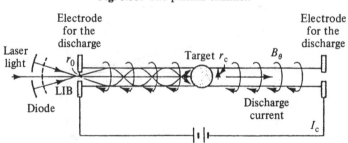

where $t$ is time. The current $J$ is given by

$$J = J_0 \exp(-Rt/2L) \sin\left(\frac{1}{CL} - \frac{R^2}{4L^2}\right)^{\frac{1}{2}} t = J_0 \exp(-Rt/2L) \sin \omega t.$$

(3.147)

The current becomes the maximum at $t = 10 \, \mu s$. When the maximum current is $I_c = 6.7 \times 10^4 \, A$, we have $\omega = 1.6 \times 10^5 \, rad/s$, $C = 8.9 \times 10^{-8} \, F$, and $J_0 = 3.0 \times 10^5 \, A$. The charging voltage $V_c$ of the capacitor bank $C$ is $V_c = L \, dJ/dt = \omega J_0 L = 2.4 \times 10^5 \, V$. Thus the energy $E_c$ stored in the capacitor bank needed to make a plasma channel is $E_c = CV^2/2 = 2.6 \times 10^3 \, J$.

If the LIB collides with electrons and ions when it passes through the plasma channel, the beam energy absorbed in the target decreases. From the point of view of the LIB's collisions, the pressure of the gas in the reactor cavity needs to be small. When neon gas of 10 Torr fills the reactor cavity, the number density $n_c$ of the gas is $n_c = 3.6 \times 10^{23} \, m^{-3}$. When a proton of 10 MeV passes through this plasma channel, the proton loses 43 keV of energy in the channel

**Fig. 3.54.** The fusion power reactor and plasma channel.

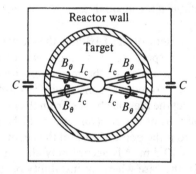

**Fig. 3.55.** A discharge circuit including the plasma channel.

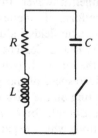

through collisions. This energy loss corresponds to a loss of 36 kJ of the total beam energy of 8 MJ, which is within the allowable order of magnitude. Since the beam current $I_b$ is much larger than the discharge current $I_c$, the beam current increases the magnetic field $B_\theta$. As a result, an electric field $-E_z$ is induced in the direction opposite to the propagation. This electric field accelerates electrons in the plasma channel to form the back current. When the number density $n_e$ of the electrons in the plasma channel is $n_e = 3.6 \times 10^{17}$ cm$^{-3}$, the back current completely cancels the beam current. Thus the beam current is neutralised by the back current as well as by electrons in the channel.

When the LIB and the back current propagate in the plasma channel, there is the possibility that the two-stream instability of the Weibel instability (usually called the filamentation instability) will occur in the channel. Suppose the number density $n_b$ of the LIB is $n_b = 7 \times 10^{20}$ m$^{-3}$. The fluctuation of the incident angles of the beam to the plasma channel determines the beam temperature. For the case where the maximum incident angle is $\varphi = 0.13$ rad, the corresponding LIB temperature $T_b = V\varphi/2 = 0.65$ MeV. This high temperature of the beam weakens the instabilities. For an LIB with such a high temperature, the two-stream and filamentation instabilities are stable if the inequalities $n_b < 10^{22}$ m$^{-3}$ and $n_c > 10n_b$ are satisfied.

Sometimes the system of the LIB and the electron back current is unstable with respect to ion sound-wave instabilities. If $n_c > 3.6 \times 10^{23}$ m$^{-3}$ is satisfied, the mean velocity of the electron flow is small and ion sound-wave instability is stable.

In order to protect the plasma channel from the micro-instabilities, the number density $n_c$ of the electrons needs to be large, although $n_c$ must be small from the point of view of the beam collisions. Neon gas with $n_c = 3.6 \times 10^{23}$ m$^{-3}$ is accordingly chosen from these two considerations. In the case where an instability occurs, it is usually electrons that play the central role, so the frequency of the instability is very high. In the channel, however, the ion beam conveys the energy to the target in the form of kinetic energy. The effect of the high-frequency electromagnetic fields induced by electrons on the beam ion is thought to be negligible. The main difficulty in beam propagation comes from the ion mode (with ion plasma frequency). But here we must note in respect of the plasma channel that the Lorentz force due to $B_\theta$ acting on the back current is in the outward radial direction. Thus the Lorentz force induces an electric field in the outward radial direction, which expands the beam ion. If the back current completely cancels out the beam current, the net force exerted by $B_\theta$ on the LIB is zero. Thus the plasma channel meets the great objection to beam transportation.

Here, a simpler and more reasonable propagation method for the LIB will be explained. From the diode shown in Fig. 3.51, the LIB is extracted, rotating around the propagation axis shown in Fig. 3.56. Imagine the reactor cavity is filled by argon at 0.1 Torr. The number density of $n_c$ of the argon is then $3.6 \times 10^{21}$ m$^{-3}$. If we assume that the temperature of the gas is 10 eV, then the gas will be ionised with unit charge. The plasma frequency $\omega_p$ for a number density $n_e = 3.6 \times 10^{21}$ m$^{-3}$ of electrons in the gas is $\omega_p = 3.4 \times 10^{12}$ rad/s. The charge of the LIB in the gas is neutralised during a time interval equal to $2\pi/\omega_p$, that is, $\tau_p = 1.9 \times 10^{-12}$ s, which is very short in comparison with the time interval of the beam propagation $\tau_b = R_r/V_b = 2.3 \times 10^{-7}$ s, where $R_r$ is the radius of the reactor ($R_r = 5$ m) and $V_b$ is the velocity of the propagation of a proton with $V = 10$ MeV ($V_b = 4.4 \times 10^7$ m s$^{-1}$). We can consider the charge to be neutralised when the LIB propagates in the reactor cavity. But the electron Larmor radius when the electron temperature $T_e = 10$ eV, under an intensity of magnetic flux density of $B = 30$ T, is $r_e = 2.5 \times 10^{-7}$ m while the mean free path $\lambda_e$ of an electron with respect to collision with ions is $\lambda_e = 3.3 \times 10^{-4}$ m. The mean Larmor radius is much shorter than the mean free path. Therefore the electron in the gas cannot follow the beam motion trapped by the magnetic field $B$. Thus the current of LIB is not neutralised in the rarefied gas. The current neutralisation factor $f$ depends on the beam current $I_b$ and the number density $n_e$ of the electrons.

Now consider the propagation of an ion beam which is extracted from the diode shown by Fig. 3.51. The non-rotating beam in the axial part is assumed to be passed through the dense plasma before

Fig. 3.56. Propagation of the rotating beam.

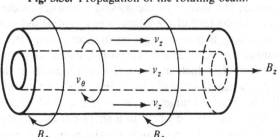

it is combined with the rotating beam. This part of the beam is then considered to be completely current-neutralised. The rotating part of the beam has a current $I_z = 1.9$ MA in the propagation $(-z)$ direction and the current $I_\theta = 1.6$ MA in the azimuthal $(-\theta)$ direction. If the current neutralisation factor is $f = 0.5$ for both directions, the net current is 1.0 MA in the $z$-direction and 0.8 MA in the $\theta$-direction. The net current $I'_z = 1$ MA in the $z$-direction induces a magnetic field $B_\theta$ in the azimuthal direction, whose maximum value reaches $B_{\theta m} = 36$ T, which confines the beam in a small radius of $r_c = 5.5$ mm. This magnetically-confined beam is unstable with respect to sausage-type instabilities unless it has a strong axial magnetic field corresponding to the toroidal magnetic field for the Tokomak (cf. section 2.1.3). The stability condition is $(2B_{zm})^{\frac{1}{2}} > B_{\theta m}$, given by eq. (2.40), where $B_z$ is the magnetic field in the $z$-direction and the suffix m refers to the maximum value. The rotating beam shown in Fig. 3.56 induces $B_z$ which has a maximum $B_{zm}$ at $r = 0$. The net current $I'_\theta = 0.8$ MA gives rise to $B_{zm} = 29$ T. These values of $B_{\theta m} = 36$ T and $B_{zm} = 29$ T satisfy the inequality $(2B_{zm})^{\frac{1}{2}} > B_{\theta m}$. The rotating beam must also be stable with respect to the kink-type instability. In the case of the ion beam, the beam propagates with a high velocity, relative to which the phase velocity of the fluctuation generally has a large velocity, and the growth rate of the fluctuation reduces. The strong magnetic fields also decrease the two-stream and filamentation instabilities. Here the propagation method by using the self-induced magnetic field is repeated. When the gas pressure in the reactor cavity is as low as 0.1 Torr, a beam of 10 MeV deposits about $10^{-6}$ of its energy in the gas of the beam path, so that the gas temperature reaches about 10 eV, at which the gas is ionised. With a magnetic field of the 30 T, the electrons in the reactor gas have a much shorter mean Larmor radius than their mean free path. Thus they are trapped by the magnetic field, and the back current of electrons to the beam current is suppressed. The charge of the beam is therefore neutralised by electrons (over the period given by the inverse of the plasma frequency of the reactor plasma) but the current of the beam is not neutralised completely by these electrons. Beam propagation induces a magnetic field $B_\theta$, which confines the beam in a small radius. Beam rotation induces a magnetic field $B_z$, which stabilises the propagation of the beam. When the multi-beams are projected at the target, the beams overlap with each other near the target, as shown in Fig. 3.57. The magnetic field $B_\theta$ which 'confines the beam' in a small radius cancels itself in the overlapped part, so it is possible for the beam to expand there.

If the number of beams is six (see Fig. 3.58(a)), the beams are at right-angles to each other and the interference of the magnetic fields becomes weak. Via an aluminium wire (see Fig. 3.58(b)) a biased voltage of −1 MV, supplied by a power-supply system of type 3 in Table 4.3, is

**Fig. 3.57.** Overlapping of beams near the target.

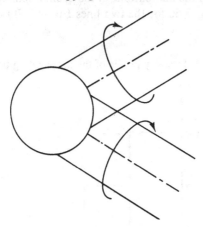

**Fig. 3.58.** The target and beam irradiation.

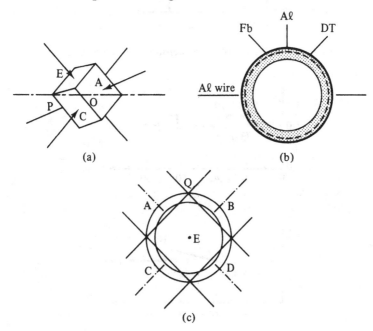

(a)

Aℓ wire

(b)

(c)

applied to the target. This negative potential of the target with respect to the ion beam, which has the potential of the earthing level since the beam passes through the diode's cathode which is itself at the earthing level, assists in focusing the beam on the target. For the case of six-beam irradiation of a target of the radius of 6 mm, if the intensity profile of the beam during propagation is that shown in Fig. 3.59, then the fluctuations of the beam intensity on the target are shown in Fig. 3.60 for the two lines EB and EQ in Fig. 3.58(c).

**Fig. 3.59.** Intensity profile of the propagating beam.

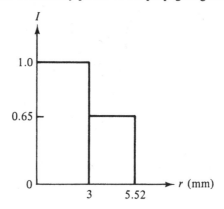

**Fig. 3.60.** Fluctuation of beam intensity on the target surface.

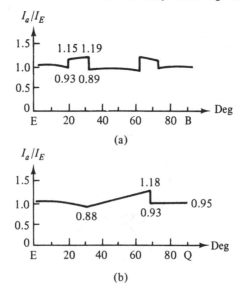

(a)

(b)

The maximum intensity fluctuation is 19 % and the average (root mean square) fluctuation is less than 5 %.

### 3.6.5. LIB interaction with the target

The target falls from the top of the reactor wall toward the reactor centre. When it falls naturally in the reactor cavity, it takes 1 s. To operate the reactor with a repetition rate of 1 Hz, the target must be pushed down with a greater acceleration. When it arrives at the reactor centre, LIB are irradiated onto the target. Since the propagation time of the LIB in the reactor is of the order of 100 ns, the displacement of the target during this time is negligible and the target seems to be at rest.

The target consists of three layers: lead, aluminium, and then the D–T fuel. To get effective implosion of the target, the central part of it is void, so that the solid D–T fuel is fixed in a thin layer inside the cryogenic aluminium layer. Thus the targets are stored out of the reactor, where they are cooled by the liquid helium. If the fuel in the target is thought to have melted or vaporised during its passage through the reactor gas, the outside of the lead layer is surrounded by a solid hydrogen layer to ensure that the fuel remains solid. When the LIB impinges on the target, the beam ions lose their energy by Coulomb scattering through interaction with mainly bound electrons of atoms in the target material. The stopping power $dE_b/dx$ of a material with density $\rho$ is given by the Bethe equation,

$$-\frac{dE_b}{\rho\, dx} = \frac{0.30708}{\beta^2}\, Z_b^2\, \frac{Z}{A}\left[f(\beta) - \log \bar{I} - \sum_i \frac{C_i}{Z} - \frac{\delta}{2}\right]. \quad (3.148)$$

Here $\beta = v/c$ ($v$ is the velocity of the beam ion and $c$ the velocity of light), $Z_b$ is the charge number of the beam ion, $A$ the atomic number of the target atom, $Z$ the charge number of the target atom, $\bar{I}$ the average ionisation potential of the target atom, $C_i$ the correction factor for a bound electron on the $i$th level, $\delta$ the correction factor for the high-energy beam, and $f(\beta)$ is given by

$$f(\beta) = \log \frac{2m_e c^2 \beta^2}{1 - \beta^2} - \beta^2. \quad (3.149)$$

Equation (3.148) gives the stopping power of bound electrons in the target atoms for beam ions due to Coulomb interactions which do not differ greatly if the state of the target material is solid, gaseous or a plasma. If we compare eq. (3.148) with eq. (3.137), which gives the stopping power of the plasma, the value given by eq. (3.148) is smaller than that given by eq. (3.137) for the same parameters. After the target is heated by the LIB, the temperature of the target materials is raised to the order of 100 eV, which ionises them. Thus the free electrons stop the beam ion more strongly than the bound electrons; the stopping range of the beam ion decreases by about 20 % when the temperature is 100 eV. The stopping power $-dE_b/dx$ in eq. (3.148) is inversely proportional to $v^2$. Therefore the beam ion loses a large part of its energy when the velocity becomes small. Figure 3.61 shows the stopping power of Pb for a proton beam which has an initial energy of 10 MeV. The stopping power $dE_b/dx$ diverges when $v = 0$; the stopping power is said to have a Bragg peak.

Figure 3.62 shows the stopping power of the Pb layer (thickness of 180 $\mu$m), and the Al layer (thickness of 320 $\mu$m) for a proton beam incident normally to the target surface with an energy of 10 MeV. The beam loses 20 % of its energy in the Pb layer and 80 % in the Al layer. If the beam energy is 8 MJ, 1.6 MJ is absorbed in the Pb layer and the remaining 6.4 MJ in the Al layer. The Al is ionised, but the temperature stops increasing at about 200 eV because of expansion.

When the LIB impinges on the target surface, each beam ion has an incident angle due to the path line and the curvature of the target surface. Thus the path of the beam ion to the target is oblique to the surface. The thickness of the stopping position of the obliquely-incident ion from the target surface is shorter in comparison with that of the normal incidence shown in Fig. 3.62.

**Fig. 3.61.** Stopping power of Pb for a proton beam of 10 MeV.

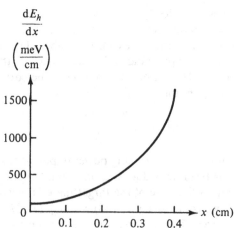

When averaged for the total beam, the Bragg peak in Fig. 3.62 disappears and the curve of the stopping power $dE_b/dx$ is smoothed out in the Al layer.

The initial densities of the target materials are $\rho_{Pb} = 11.3$ g cm$^{-3}$; $\rho_{Al} = 2.79$ g cm$^{-3}$; and $\rho_{DT} = 0.21$ g cm$^{-3}$. If the initial thicknesses of the layers are $\delta_{Pb} = 18$ $\mu$m, $\delta_{Al} = 320$ $\mu$m and $\delta_{DT} = 200$ $\mu$m, then the mass ratios are $M_{Pb}:M_{Al}:M_{DT} = 49:22:1$. The centre of mass is located in the Pb layer. Initially, the Al density is 13 times greater than the D–T density. The boundary between the fuel and the pusher is safe from Rayleigh–Taylor instability given the strong acceleration toward the centre. When the fuel approaches the target centre, the Al layer expands about 40 times in volume. The Al density at this stage is lower than that of the fuel density. Thus the boundary is again stable with respect to Rayleigh–Taylor instability in the deceleration phase of the implosion. The target materials Pb and Al have small induced radio-activities.

There are two main reasons for choosing Pb and Al as the target materials. The first is that they have appropriate densities for serving as the tamper and the pusher, the second that they have small induced radio-activities. The Al layer expands, absorbing the beam energy. The Al layer expands exclusively inward, accelerating the D–T fuel to the target centre, because the heavy Pb layer surrounds the outside of the Al layer. A proton beam with 4 MeV has a good stopping range in the metal layers, and the beam energy absorbed in the

**Fig. 3.62.** Stopping power of Pb and Al layers in the target for a proton beam of 10 MeV.

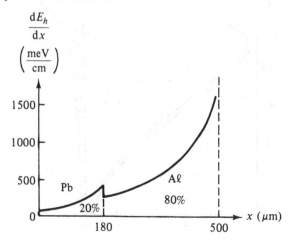

target effectively accelerates the fuel, rather than being spent in blowing off. By comparison with the case in which the deflagration wave is formed on the target surface, LIB interaction with the target leads to target implosion with a high hydrodynamic efficiency.

Figure 3.63 shows the result of the simulation of target implosion in the $r$–$t$ diagram. The figure shows that the Al layer expands. But the Pb layer expands too, pushing against the Al layer. Thus the boundary between the Pb and the Al layers remains constant during the implosion. The Pb layer thus plays the role of a tamper efficiently. The Al layer expands enough to convert almost all its thermal energy to kinetic energy of the D–T and the Al layers.

The pressure of the Al layer reaches $10^{13}$ Pa (at a temperature of 200 eV) and pushes the D–T layer with an acceleration of $10^{12}$ m s$^{-2}$. The implosion velocity reaches $2 \times 10^5$ m s$^{-1}$. When the fuel layer approaches the target centre, the cross-sectional area of the fuel path becomes narrow and the fuel is decelerated rapidly. The momentum of the fuel together with the Al layer is changed into an impulse. Since the time interval of the deceleration is very short, the pressure inside the fuel layer increases to $10^{18}$ Pa. This high pressure compresses the fuel to $\rho R = 7.0$ g cm$^{-2}$ and heats the fuel to an ion temperature of $T_i = 4$ keV. The burn fraction of the fuel will reach 34 % and the target will release fusion energy of 2.5 GJ. For an input energy of 8 MJ from the six beams, the target gain is $G = 310$.

**Fig. 3.63.** The $r$–$t$ diagram for target implosion.

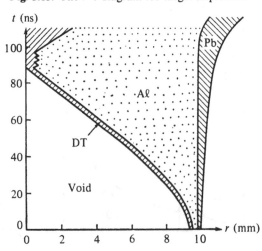

In order for the fuel to achieve a high $T_i$ and a high $\rho R$, spherically-symmetric implosion is required. The deposition of the energy of the six beams in the pusher layer has a fluctuation of 4 %. This induces inhomogeneity of pressure in the pusher layer, and the thermal conduction of the pusher layer is large enough to smooth out this fluctuation. Rather, the radiative energy transfer plays this role. The intensity of the radiation is proportional to $Z^{\frac{7}{2}}$. If a little high-$Z$ material is mixed with Al, the radiation intensity in the pusher layer can increase without large increase in density.

The fuel density during the implosion is high while its temperature is rather low. The electrons in the fuel are Fermi-degenerated, while the deuterium ions follow the Bose–Einstein statistics and are not degenerated. But the coupling coefficient $\Gamma = Z^2 e^2/4\pi\varepsilon akT$ of the plasma, which is the ratio of the Coulomb potential $Z^2 e^2/4\pi\varepsilon a$ to the thermal energy $kT$, exceed 0.01 (weak-coupling plasma), stemming from the small mean ion distance $a$ which is due to the high density. The transport coefficients, especially the equation of state of the weak coupling plasma with electron degeneracy, allow large modifications compared to those of an ideal plasma. We must take this fact into account in calculations. In summary, we can say that the mechanism of LIB fusion is becoming clear and the construction cost of the LIB accelerator is reasonable. LIB fusion looks like being the most promising method in the near future.

## 3.7. The heavy-ion beam and projectile

### 3.7.1. The heavy-ion beam

There are proposals to use the heavy-ion beam (HIB) as the energy driver. When $^{238}_{92}$U with an energy of 10 GeV is projected at the target, the energy per nucleon is 42 MeV.

The bound electrons of U are stripped in the target metal when the target moves with a large velocity and has a high charge. Thus the stopping range of U in the target is comparable to H with an energy of 4 MeV. (For the HIB, where the charge of the target is very important and determines the stopping range, the dependence of this effective charge of the HIB on the material temperature at present seems somewhat ambiguous.) The HIB energy per particle is 1000 times that for the LIB, and the beam current of the HIB reduces to 1/1000 the current of the LIB. This is the great advantage of the HIB. Since the particle energy of the HIB is much larger than that of the LIB or the REB, the HIB cannot be accelerated by a

Marx generator, and the advantage of the low cost of the accelerator for the LIB or the REB disappears for the HIB.

One charge of the HIB is accelerated by a linear accelerator (linac). An HIB with one charge from the ion source, which consists of the discharge tube and the Cockcroft–Wolton type of accelerator, is sent to the Vider̈oe accelerators through the RFQ (radio-frequency quadrupole accelerator). The Vider̈oe accelerator shown in Fig. 3.64 accelerates the ion successively in the gaps between the electrodes through the action of alternating voltage. To increase the beam current, two beams are combined into a single beam several times, as shown in Fig. 3.65.

The HIB accelerated to 30 MeV by the Vider̈oe accelerators is sent to the Albarez accelerators and is then accelerated to 10 GeV. The total length of the linacs reaches 10 km. The HIB is overlapped in the accumulation ring (radius 100–1000 m) to increase the current

**Fig. 3.64.** The Vider̈oe-type accelerator.

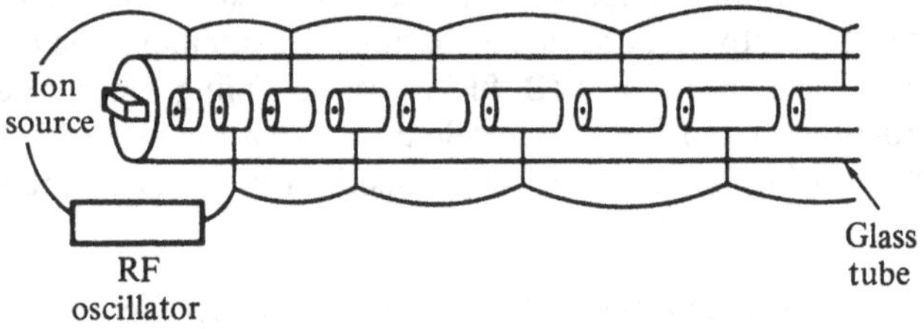

Ion source

RF oscillator

Glass tube

**Fig. 3.65.** The HIB accelerator system.

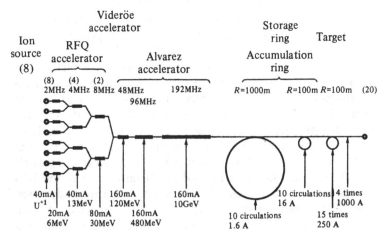

and is sent to the storage ring, where the beam length is shortened to increase the current density.

The HIB accelerated by linacs has a high emittance due to the small divergence angle, and ballistic projection to the target through a hole of the reactor wall is possible. The charge of the beam is not neutralised in the accelerator, so the beam has a space-charge limit. Due to the low thermal spread, the HIB may suffer filamentation instability, together with other types of instability, especially in the storage ring. The beam current must be suppressed to the order of 1 kA. For the case where 20 kA of HIB current is projected at the target, the HIB comprises 20 beams.

After the accumulation ring, the HIB is split into the 20 beams in the 20 storage rings, and these 20 beams are projected into the reactor cavity simultaneously.

The target for the HIB is quite similar to that for the LIB. The energy conversion rate from electrical energy to the HIB kinetic energy is 10 %, and can rise to 20–25 %. The linac systems for nuclear experiments have so far operated steadily. But the currents of the beams in the conventional accelerators are of the order of $\mu$A or mA, so there are some problems in obtaining the large current of 1 kA required by the accelerators, for such a strong current is very expensive. It is thus necessary to develop an accelerator for a large current that nevertheless has a low construction cost. The induction linac is one candidate for such an accelerator.

### 3.7.2. The projectile

As the energy driver for inertial-confinement fusion, a heavier mass (laser → REB → LIB → HIB) has been proposed. Instead of an ion, a projectile of a mass of several mg has recently been suggested (Fig. 3.66). When a strong current flows along the rail gun, the

**Fig. 3.66.** Rail gun and projectile.

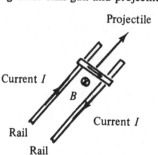

magnetic field accelerates the projectile. If the target is hit by several projectiles simultaneously, the fuel in the target will release fusion energy. Compared to the required velocity of $10^5 \, \mathrm{m \, s^{-1}}$ for the projectile, achieved velocities now are one or two orders of magnitude too small. A method of compressing the fuel in a spherically-symmetric way must also be developed.

# 4

# The fusion reactor

•

*The fusion power reactor is to be constructed by concentrating on wholly new and extremely sophisticated technology. What kind of technology is required? What special characteristics does the fusion reactor have? This chapter seeks answers to these questions.*

## 4.1. The structure of the fusion power reactor

### 4.1.1. The concept of the fusion power reactor

The fusion power reactor converts the energy released by the fusion reaction to the form in which we want to use it. The design of the reactor is conceived in accordance with the method used to confine the plasma.

To begin this chapter, the various reactors are classified. The conceptual design of a reactor with the magnetically-confined plasma is then described, and the elements of the reactor are examined from the point of view of the new technologies they involve. The Tokomak and the mirror reactor are looked at in detail. Next, reactors with inertially-confined plasma are described, notably for the laser and the light-ion beam types. Finally, the fuel resources required to operate the fusion power reactor are investigated, and the effect on the environment and safety of the reactor are considered.

The objective of fusion research is to construct a fusion power reactor and to make the energy of fusion available for practical purposes. The main task of fusion research for the past 30 years is to find a stable way of confining plasma at high temperature. Recently, fusion power reactors have been extensively re-designed to keep up with developments in research of the confinement of plasma, although no plasma yet satisfies the Lawson criterion.

The conceptual design of the fusion power reactor is clearly important if we are to develop the technologies required to construct the reactor, to detect the gaps between present and required technological levels, and to estimate the time it will take to select

Table 4.1 *Classification of types of reactor according to the method of plasma confinement*

| Method of confinement | Configuration of magnetic field | Facility | Operation mode | Chapter section for explanation |
|---|---|---|---|---|
| Magnetic confinement | Torus | Tokomak | Quasi-steady | 2.2 |
| | | Reverse field pinch | Quasi-steady | 2.3.1 |
| | | Stellarator | Steady | 2.3.2 |
| | | Heliotron | Steady | 2.3.2 |
| | | Torus pinch | Pulse | 2.3.8 |
| | Open end | Mirror | Steady | 2.3.3–5 |
| | | Straight pinch | Pulse | 2.3.8 |
| | | Plasma focus | Pulse | 2.3.10 |
| Inertial confinement | | Laser | Pulse | 3.2, 3.3, 3.4 |
| | | REB | Pulse | 3.5 |
| | | LIB | Pulse | 3.6 |
| | | HIB | Pulse | 3.7.1 |
| | | Projectile | Pulse | 3.7.2 |

and develop the best technology, and to seek the funds and the manpower needed by this technology.

Let us now classify reactor types according to the methods of confining the plasma (Table 4.1).

The reactors are first divided into two groups: magnetic confinement and inertial confinement. Because the structure of the two groups of reactors do not differ much, in this section descriptions are given only for the magnetic-confinement group, to avoid confusion. Reactors using inertial confinement will be described in section 4.3.

Section 1.3 described the Lawson criterion. Since the cross-sectional area of fusion for the D–T reaction is larger than for the other type of fusion reactions, in addition to the fact that a lower temperature is required for the reaction, fusion power reactors of the first generation will be D–T reactors. In this section description is limited to the D–T reactor only.

Fusion power reactors with magnetically-confined plasma may be divided into the torus type and the open-end type, depending on whether the closed magnetic fields completely cover the plasma surface or whether the magnetic fields are partly open.

According to the operation mode, the fusion power reactors may also be classified into the steady, the quasi-steady and the pulsed operation types.

   (i) *Steady reactor:* In these reactors, fusion reactions continue steadily. Reactors using the stellarator, heliotron and the mirror belong to this type of reactor.

  (ii) *Pulsed reactor:* The reactor is stopped and started repeatedly. The burning time is of the same order as the confinement time of the plasma. (In talk about fusion reactors, the word 'burning' is frequently used for the fusion reactions.) Reactors using the z-pinch, the theta-pinch and the plasma focus belong to this type.

 (iii) *Quasi-steady reactor:* This is an intermediate type of reactor compared to those described above. The unsteady burning is designed to continue for a much longer period than the confinement time of the plasma. However, the period of burning does have a limit. A typical example of this type is the Tokomak.

The general features of the D–T power reactor for magnetic-confinement fusion is shown in Fig. 4.1. At the reactor centre is the high-temperature plasma, which consists of 50 % D and 50 % T.

**Fig. 4.1.** Constitution of the fusion power reactor.

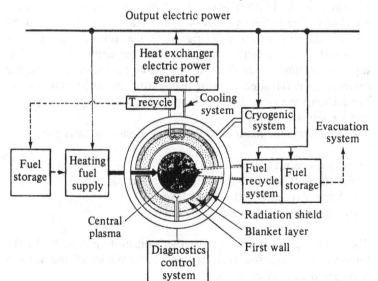

Since the high-temperature plasma must avoid the least contamination, the plasma is confined by the magnetic field in a vacuum inside the vessel.

The vessel wall facing the plasma is called the first wall. Those plasma particles with a high temperature diffuse from the main body of the plasma confined at the reactor centre and hit the first wall. The plasma particles can damage the first wall, which shortens the life of the first wall and contaminates the plasma. To limit damage to the first wall by plasma particles, the reactor has a diverter or gas blanket.

The $\alpha$-particle, a product of the D–T reaction, has a particle energy of 3.5 MeV. Since the $\alpha$-particle is an ionised nucleus, it is confined by the magnetic field, losing its energy by collision with the plasma and diffusing out of the main plasma. The neutron, the other fusion product, passes through the first wall because it is charge-neutral and has a large particle energy of 14.1 MeV, and impinges on the blanket behind the first wall. This blanket converts the kinetic energy of the neutron into thermal energy, as well as breeding tritium in it to provide the fuel for the reactor.

The blanket is surrounded by the shield layer, which prevents outflow of neutrons and $\gamma$-rays. The superconducting magnet coil outside the shield layer confines the plasma to the central part of the reactor.

The reactor has a cooling system, by which the thermal energy absorbed in the blanket is conveyed by the coolant to the heat exchanger. The tritium produced in the blanket, and the unburnt tritium and deuterium diffusing from the central plasma, are separated and collected by the fuel recycling system. The superconducting magnet is cooled by the cryogenic system, and the central plasma is heated to fusion temperature by the heating system. In addition, the reactor has a fuel supply system, a diagnostic system, and a control system.

In summary, the main parts of fusion reactor systems are:

   (i) the central plasma;
   (ii) the first wall and the structural materials;
   (iii) the blanket and the shield layer;
   (iv) the superconducting magnet coils.

These main parts connected together constitute a reactor. In the following sections, the functions and technologies of the various parts are described in detail.

### *4.1.2. The central plasma*

Plasma at fusion temperature is located at the centre of the reactor. The D and the T ions are confined in the central part during the confinement time, on average, and leave there by the diffusion. The plasma density at the centre must be held constant, in the steady or the quasi-steady reactor, by supplying an amount of fuel corresponding to what is lost by diffusion. This fuel must be supplied to the reactor centre, where it must be heated to fusion temperature, while the unburnt fuel diffusing out from the main plasma must be separated and recycled.

### *Heating the plasma*

Typical heating methods such as Joule heating, neutral-beam injection, and wave heating were described in sections 2.2–2.4. The self-heating of the plasma by $\alpha$-particles, which are one of the fusion products and have a particle energy of 3.5 MeV, is called the $\alpha$-particle heating or internal heating, and was described in section 1.3.1. Here, $\alpha$-particle heating is considered once more in connection with the economy of reactor operation.

Electric power must be supplied to the reactor to heat the cold fuel and to compensate for the energy loss, e.g. by radiation. The ratio of fusion power $P_f$ to the externally-supplied power $P_h$ is denoted by $Q$, i.e.

$$Q = \frac{P_f}{P_h}. \tag{4.1}$$

Since $P_f$ is the fusion energy released in the plasma per unit time, it has the value $P_f = \langle \sigma v \rangle n^2 E_f V/4$ given by eq. (1.33), where $V$ is the volume.

The fact that $Q$ is large means that the externally-supplied power is small. Thus $Q$ is directly related to the economy of the fusion power reactor. When the reactor is operated steadily, supported only by $\alpha$-particle heating, externally-supplied power is required only to start the reactor. In steady operation of the reactor, $P_h \to 0$ and $Q \to \infty$. The condition for steady operation is given by eq. (1.46) in section 1.3.4.

### *Supply of fuel*

The methods of supplying the fuel to the central plasma are:

  (i) injection of a neutral beam, or cluster, or pellet of the fuel;
 (ii) diffusion of the fuel in a gas blanket.

Neutral-beam injection (NBI) described in section 2.2.3 supplies
fuel and heats the plasma. This method is applied in the open-end
system, in which the $Q$ value is small and power is continuously
supplied externally. The cluster, secondly, consists of $10^1$–$10^9$
molecules. When a cluster with unit charge is accelerated by the
electric field, it reaches the deep region of the plasma, not being
affected much by the magnetic field since the ratio $e/m$ is very small.
In the case of the pellet injection, the fuel in the pellet, whose radius
is several mm, is supplied to the central plasma. When the pellet
advances into the plasma, the pellet surface evaporates, ionising in
the plasma, but the core fuel of the pellet reaches the centre of the
plasma. The pellet size must be chosen carefully to match the reactor
radius and the injection speed.

In the gas-blanket method the neutral fuel gas fills the gap between
the central plasma and the first wall. This neutral gas supplies fuel
to the plasma by diffusion. The neutral fuel gas also serves to protect
the first wall, to suppress contamination of the plasma and to treat
the particles diffusing from the central plasma.

It has been proposed that the high pressure of the neutral gas
should itself confine the plasma at fusion temperature. Many other
proposals, however, aim to use neutral gas with a much lower
pressure than the pressure of the central plasma, which in such cases
is confined by a magnetic field. The ionised particles in the central
plasma cannot cross the magnetic field, while the neutral fuel gas
can diffuse into the plasma across the magnetic field. Thus the fuel
is automatically supplied to the plasma.

The $\alpha$-particle, as the fusion product, and the unburnt
high-temperature D and T ions leave the central plasma by diffusion.
Before they hit the first wall, they collide with the neutral gas, losing
their energy and combining with electrons to become neutralised,
and are then evacuated from the reactor cavity. Thus the neutral gas
protects the first wall from damage through bombardment by
high-speed ions.

As mentioned above, the gas blanket offers the possibility of
helping to solve some of the difficulties associated with
magnetically-confined plasma. Because of the uncertainty of the
effect of the gas blanket on the confinement of plasma and the cooling
of the fusion temperature, however, practical applications are yet
unrealised and so remain a topic for theoretical and experimental
investigations in the future.

## The particle balance and burn fraction

If the injection rate of the fuel (particles of a mixture of D and T specified per $m^3$ and per s) is denoted by $S$, the temporal change $dn/dt$ of the number density of the plasma is given by

$$\frac{dn}{dt} = S - \frac{n}{\tau} - 2\left(\frac{n^2}{4}\langle\sigma v\rangle\right), \tag{4.2}$$

where the number density of D is assumed to be equal to that of T, i.e. $n_D = n_T = n/2$. The second term on the right-hand side of eq. (4.2) is the number of particles lost by the diffusion per unit time, where $\tau$ is the confinement time. The third term gives the plasma particles lost by the fusion reactions. Two particles are lost per reaction. In the steady state $dn/dt = 0$, we have

$$S = \frac{n}{\tau} + \frac{n^2\langle\sigma v\rangle}{2}. \tag{4.3}$$

During experiments with recent Tokomaks, we have found that the particles which are back-scattered after hitting the first wall, and the particles which are re-emitted after being absorbed once in the first wall, subsequently enter the central plasma and increase the plasma density, perhaps disturbing the particle balance. This phenomenon is called recycling of plasma particles. It will be investigated in detail in the future if the size of the reactor and the plasma density increase greatly.

Some of the plasma leaves the central plasma unburnt via diffusion. The burn fraction $f$ of the fuel is defined by

$$f = \frac{n^2\langle\sigma v\rangle/2}{S}. \tag{4.4}$$

If $S$ from eq. (4.3) is substituted into eq. (4.4), we have

$$f = \frac{n\tau\langle\sigma v\rangle}{2 + n\tau\langle\sigma v\rangle}, \tag{4.5}$$

where $\langle\sigma v\rangle$ is a function of the temperature and is given in Fig. 1.4. For a temperature of 20 keV for the D–T plasma, we have $\langle\sigma v\rangle = 4 \times 10^{22}\ m^3\ s^{-1}$. If we choose the values $1 \times 10^{20}\ s\ m^{-3}$ and $1 \times 10^{20}\ s\ m^{-3}$ for $n\tau$, the burn fraction $f$ is,

    (i) $f = 0.020$ for $n\tau = 1 \times 10^{20}\ s\ m^{-3}$;
    (ii) $f = 0.091$ for $n\tau = 5 \times 10^{20}\ s\ m^{-3}$.

Generally, the burn fraction is several per cent.

The unburnt plasma particles are evacuated from the reactor cavity. With a small $f$, there is low economy in the operation of the fusion reactor, due to the low utilisation of the heated plasma. The D component of fuel can be obtained easily and safely at a low price, but T is a radio-active material which does not exist in nature, and must be bred in the gas blanket. The T must be recycled completely, and not allowed to leak out into the atmosphere.

### 4.1.3. The first wall

Among the structural components of the fusion power reactor, the one which is directly exposed to the central plasma is called the first wall. Since the first wall will be exposed to the most severe conditions in the reactor, the technical problems it presents are examined first here.

The first wall is irradiated by the unburnt fuel, D and T; the nuclear fusion products, $\alpha$-particles and neutrons; contaminating atoms; and electromagnetic waves, X-rays and $\gamma$-rays. Behind and surrounding the first wall are blanket materials (lithium metal or molten lithium salts), which operate at a high temperature and are strongly corrosive. And finally, the first wall must survive repeated thermal stress, in the case of reactors of the pulse-operation or of the quasi-steady type. These are the mechanical, chemical, thermal, and nuclear characteristics which must be taken into consideration when we select the material for the first wall.

### The wall loading

The wall loading refers to the various energy fluxes described above which impinge on the first wall per unit area and per unit time. If we are considering the neutron flux only, we speak of the neutron wall loading. The wall loading is a measure of the extreme conditions the first wall endures in facing the plasma. When the wall loading is large, radiation damage to the wall is severe and the lifetime of the wall becomes short. But if the wall loading is kept small the reactor becomes large, because the area of the wall is then larger for the same output of the reactor. The choice of the value for the wall loading is directly related to the economy of the reactor. In the preliminary stage of the conceptual design of the reactor, a large value such as $10 \, \mathrm{MW \, m^{-2}}$ was chosen for the wall loading. However, the value was then lowered after the severe radiation damage done to the wall by the strong energy flux through it was appreciated. At present, $1–3 \, \mathrm{MW \, m^{-2}}$ is considered to be reasonable for the wall loading. The radiation damage to the first wall and plans to counter it will now be examined; the damage will be divided into that done by atoms and that done by neutrons.

### Wall damage by atoms

The most important damage to the wall involves sputtering and blistering. The phenomenon of atoms colliding with the surface of the wall material and knocking out atoms from the surface layer is called sputtering. Research on sputtering has a long history and the results of many theoretical and experimental research projects have been published. However, these results have not agreed with each other very well, particularly when light atoms hit a metal surface with a high melting temperature (as envisaged for the wall of a fusion power reactor). Figure 4.2 shows the sputtering yields (the ratio of the number of knocked out atoms to the number of incident atoms) when the light atoms hit Nb and stainless steel. In the figure, the sputtering yield of the atoms of the wall material itself, which upon being knocked out and contaminating the plasma then impinge on the wall again, are drawn as the Nb curve. Note that it is incorrect to say that the sputtering yields have maxima in the range 2–10 keV, which is the temperature of the fusion plasma.

Let us estimate the wear and tear on the first wall by sputtering. All D atoms which leave the central plasma are assumed to hit the first wall. The incident number of the D atoms on the first wall is

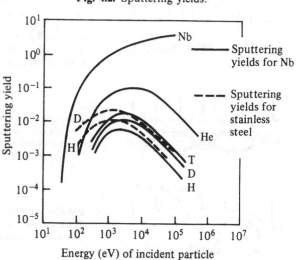

Fig. 4.2. Sputtering yields.

then about $10^{20}$ particles m$^{-2}$ s$^{-1}$. The first wall is assumed to consist of Nb. If we use the sputtering yield of $4.2 \times 10^{-3}$ for D $\rightarrow$ Nb, the exhaustion number of Nb atoms is $4.2 \times 10^{17}$ particles m$^{-2}$ s$^{-1}$. This exhaustion number corresponds to a decrease velocity of $11.2 \times 10^{-12}$ m s$^{-1}$ in the thickness of the metal surface. Supposing that the first wall must be replaced when its thickness of 6 mm has decreased 20 %, the replacement time for the first wall is 3.5 years. Since such a lifetime of the first wall is too short, there is a need to develop a new technology to protect it from the sputtering.

When the gaseous atoms impinge on a metal, the atoms penetrate the metal surface and make bubbles. These bubbles grow and eventually explode, stripping off the surface layer of the metal. This phenomenon is called blistering. Figure 4.3 shows schematically the

**Fig. 4.3.** The mechanism of blistering.

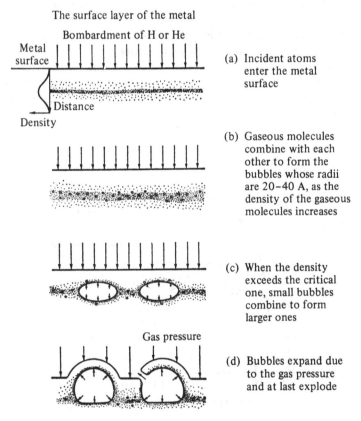

The surface layer of the metal

(a) Incident atoms enter the metal surface

(b) Gaseous molecules combine with each other to form the bubbles whose radii are 20–40 A, as the density of the gaseous molecules increases

(c) When the density exceeds the critical one, small bubbles combine to form larger ones

(d) Bubbles expand due to the gas pressure and at last explode

mechanism of blistering and Figure 4.4 the pattern of blistering that occurs when accelerated He impinges normally to the surface of Nb.

The blistering depends on:

   (i) the energy of the incident atom – this determines the distance of penetration;

**Fig. 4.4.** Blistering for $He^+$–Nb.

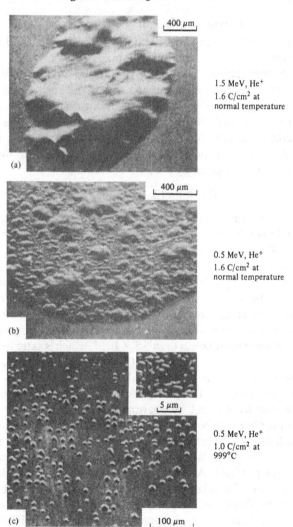

400 μm

1.5 MeV, $He^+$
1.6 $C/cm^2$ at
normal temperature

(a)

400 μm

0.5 MeV, $He^+$
1.6 $C/cm^2$ at
normal temperature

(b)

5 μm

0.5 MeV, $He^+$
1.0 $C/cm^2$ at
999°C

(c)

100 μm

(ii) the kind of atom: when the coefficient of diffusion is large, blistering scarcely occurs;
(iii) the number of incident atoms: the size of bubbles depends on the number of the incident atoms;
(iv) the temperature of the incident atoms: this determines the coefficient of diffusion.

The number of atoms ejected from the metal by blistering is estimated to be 100 times the number of atoms ejected by the sputtering.

### Plasma contamination and protection

Sputtering and blistering not only damage the first wall and shorten its life; they also contaminate the plasma.

The bremsstrahlung (cf. section 1.3.2) and cyclotron radiation induce radiation loss in the plasma. Since the energy lost by the bremsstrahlung is greater than that by cyclotron radiation when the plasma burns steadily, only the bremsstrahlung energy loss will be considered here. The ratio $R$ of bremsstrahlung from H plasma with an impurity, whose concentration is $f$ and charge is $Z$, to that from the pure plasma is given by

$$R = 1 + f(Z + Z^2) + f^2 Z^3. \qquad (4.6)$$

High-$Z$ atoms such as Fe, Mo, Ni, Nb and V which are used as the wall materials become bare nuclei at the fusion temperature of 10 keV, all the bound electrons are being stripped off. When such atoms contaminate the plasma, the energy loss by bremsstrahlung becomes quite large. If 1 % of O atoms mix with the H plasma, the energy loss by the bremsstrahlung increases 77 %. In order that the increase in the energy loss through bremsstrahlung be kept less than 10 % in a practical fusion power reactor, the concentration of Nb in the H plasma must be less than $5.8 \times 10^{-5}$, which is extremely low.

Keeping in mind this requirement, we must design the reactor so that plasma which leaves the central part by diffusion does not hit the first wall directly. The gas blanket plays the role of protecting the first wall from direct bombardment by plasma. Other devices for this purpose are the limiter and the diverter. Figure 4.5 depicts various limiters. The limiter restricts the expansion of the plasma column and so protects the first wall from direct bombardment of the plasma particles. As explained in section 2.2.2, the electrical resistance of the plasma decreases rapidly with increasing temperature. When the electric field applied to the plasma is large, the electron drift velocity becomes large also, while the electron mean free path becomes remarkably large. Therefore the electrons drift farther and faster, without experiencing collisions. Such electrons are called runaway electrons. In a Tokomak, runaway electrons are accelerated by the toroidal electric field and deviate from the main path. Thus these runaway electrons induce an electric field in a radial direction, which enhances ion diffusion across the magnetic fields. The limiter first prevents the runaway electrons from deviating from the main path and restricts the plasma radius within a narrow limit. Figure 4.6 shows how impurities around the plasma column decrease the temperature by radiative energy transfer, and how the limiter plate absorbs the impurities outside of it. Besides a metal limiter plate, a magnetic limiter using a magnetic field to confine the expansion of the plasma column is also being examined. Limiter plates operate in combination with an evacuation pump.

Figure 4.7 shows a diverter which pulls the outer part of the poloidal magnetic fields out of the plasma column by the action of a diverter coil. Plasma diffusing from the central region flows along the magnetic field lines outside the separator point. The diverter coil

**Fig. 4.5.** Various limiters.

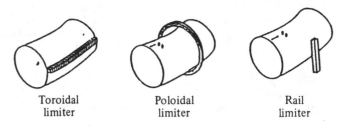

Toroidal         Poloidal          Rail
limiter          limiter          limiter

**Fig. 4.6.** Cooling of the plasma near the boundary by bremsstrahlung from impurities and the limiter.

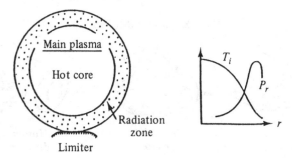

leads the outer magnetic field lines from the separator point to the outside of the plasma column. The impurities sputtered out from the first wall remain in the boundary region of the plasma column. These impurities are trapped by the diverter fields and are deposited on the diverter plate. Figure 4.8 shows the three kinds of diverter. Since the diverter plate, which is cooled by the water, must receive the severe particle (and energy) flux, its operation is assisted by the evacuation pump.

A poloidal diverter is more easily set onto a plasma column whose cross-sectional area has vertically-elongated shape. If the shape of the cross-sectional area of the plasma column approaches a circle, in the practical fusion power reactor of the future, there will be some difficulties from the point of view of stability in making the diverter operate effectively.

To reduce contamination of the plasma and to prevent damage to the first wall, it has been suggested that the first wall be coated with a low-$Z$ material. As coating materials, $B_4C$, $SiC$, $TiC$, and $C$ are being examined. However, these materials have the disadvantage

**Fig. 4.7.** The diverter.

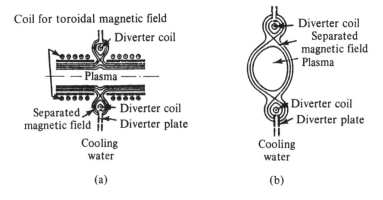

(a)                   (b)

**Fig. 4.8.** Various diverters.

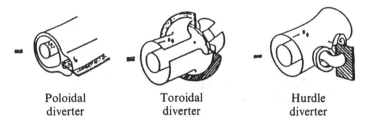

Poloidal         Toroidal         Hurdle
diverter          diverter         diverter

Table 4.2 *Operating environment of the first wall*

| Parameter | Value |
|-----------|-------|
| Temperature (°C) | 300–1000 (in steady state) |
| | $\Delta T = 50$–300 (thermal pulse) |
| Irradiation of neutrons ($n$ cm$^{-2}$ y$^{-1}$) | $10^{21}$–$10^{22}$ (14.1 MeV)* |
| | $10^{22}$–$10^{23}$ (total)* |
| Damage (dpa s$^{-1}$) | $10^{-6}$–$10^{-1}$ |
| Gas growth (ppm y$^{-1}$) | 75–500 (hydrogen)** |
| | 30–600 (helium)** |

\* For wall loading of 10 MW m$^{-2}$
\*\* For wall loading of 1 MW m$^{-2}$

that their sputtering yields are higher than those of materials with high melting temperatures. As a measure for the suitability of a material for the first wall, we may use what is called the 'critical density of the impurity in the plasma/sputtering yield'.

The first wall of the Tokomak coated by TiC includes C as well as N. Recent experiments show that repeated flashings of the TiC surface reduces contamination of the plasma by the C in the coating film of TiC on the first wall.

### Wall damage by neutrons

Damage by sputtering and blistering is limited to near the surface of the first wall. However, neutrons cause damage all over the material of the wall. The gas blanket, the limiter, the diverter, and the coating of the wall do not prevent damage from neutrons.

Damage to the wall and the structural materials of the reactor may be divided into

(i) damage caused when a neutron collides with one of the atoms constituting the solid crystal, and knocks the atom out of the crystal lattice;

(ii) damage whereby a gas is formed, or the composition of the solid material is changed.

The operating environment and the damage caused to the first wall, when exposed to the most severe neutron flux, are given in Table 4.2.

A neutron has no charge and collides with the nucleus directly without Coulomb interactions with electrons and ions. If the collided nucleus receives more energy than the threshold $E_d$, the nucleus is knocked out of the lattice (it is called the primary knock-on atom), leaving a hole there. When the energies of these primary knock-on atoms exceed the threshold $E_d$, they knock nuclei successively out of the lattice. The situation is explained in Fig. 4.9. When the atom (nucleus) moves out of the lattice, we speak of the displacement of the atom. In Table 4.2, the term dpa stands for this displacement. When the atomic holes deepen and form voids, the volume of the crystal metal increases and its mechanical strength decreases. This phenomenon is called swelling and was first observed in the stainless steel of the tube covering the fuel in the first breeding reactor. When highly fluent neutrons impinge at high velocities on a metal which has a temperature of the order $0.3–0.6T_m$, swelling appears in the metal. Here $T_m$ is the absolute temperature at which the metal melts. Such swelling is under investigation now for both the covering tube of the fuel in the breeding reactor and in the structural materials for the fusion power reactor.

As Table 4.2 indicates, the bubbles of H or He gas which grow in the first wall when irradiated by a neutron flux come from the (n, p) or (n, α) nuclear reactions. When a nucleus $X$ whose atomic number is $a$ and mass number is $b$ absorbs a neutron and undergoes a nuclear reaction, the reaction may be written as follows:

(i) (n, p) reaction:

$$^b_a X + ^1_0 n \rightarrow ^{\ b}_{a-1} Y + ^1_1 H. \qquad (4.7)$$

**Fig. 4.9.** Damage caused when a lattice atom is knocked out by a neutron.

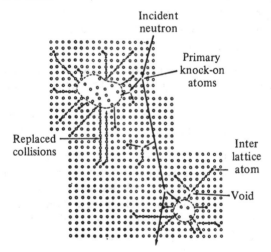

(ii) (n, α) reaction:

$$_a^b X + _0^1 n \rightarrow _{a-2}^{b-3} Z + _2^4 He. \tag{4.8}$$

Of these H and He atoms, the He atoms have a small diffusion coefficient and accumulate as voids in the first wall, causing the swelling.

It is clear from eqs (4.7) and (4.8) that the atom $X$ is changed into another atom $Y$ or $Z$ by the action of a neutron. Thus different atoms accumulate in the first wall via these nuclear reactions. New atoms can be produced through many other nuclear reactions besides eqs (4.7) and (4.8). If the new atom is radio-active, it is called an induced radio-active atom. The induced radio-activities of the wall and structural materials are very important for the safety of a fusion power reactor. This problem will be taken up again in section 4.4.2. If Nb is used as a structural material in a reactor for 20 years, Mo and Zr are estimated to accumulate to 9.5 % and 13.5 %, respectively, of the Nb, via nuclear reactions. These concentrations are clearly not small.

### 4.1.4. The blanket shield layer

The blanket serving as the shield layer has three functions:

(i) High-energy neutrons produced by the D–T reaction are slowed down and their energy is absorbed, changing the thermal energy of the coolant which is sent to the heat exchanger outside the reactor.

(ii) Tritium, which does not exist in nature, is bred.

(iii) The neutron flux and the $\gamma$-ray flux are shielded so they do not reach the superconducting magnet.

If the superconducting magnet is subjected to the radiation, its superconductivity is diminished and the temperature increases. This problem will be discussed in the next section; here let us deal with functions (i) and (ii).

### The fuel cycle

The fuel cycle for the D–T fusion power reactor may be described as follows:

(i) the D–T reaction in the reactor,

$$_1^2 D + _1^3 T \rightarrow _2^4 He + _0^1 n + 17.6 \text{ MeV}; \tag{4.9}$$

(ii) T-breeding in the blanket,

$$_3^6\text{Li} + _0^1\text{n} \rightarrow _2^4\text{He} + _1^3\text{T} + 4.8 \text{ MeV}, \qquad (4.10)$$

$$_3^7\text{Li} + _0^1\text{n} \rightarrow _2^4\text{He} + _1^3\text{T} + _0^1\text{n} - 2.5 \text{ MeV}; \qquad (4.11)$$

(iii) the reaction for neutron multiplication,

$$_4^9\text{Be} + _0^1\text{n} \rightarrow 2_2^4\text{He} + 2_0^1\text{n}. \qquad (4.12)$$

In the D–T reaction, neutrons produced by reaction (4.9) breed T in the blanket by reactions (4.10) and (4.11). In other words, the fuels for the D–T reactor are D and Li. Both $_3^6\text{Li}$ and $_3^7\text{Li}$ are stable isotopes and constitute 7.5 % and 92.5 %, respectively, of mined lithium. Reaction (4.10) is exothermic while reaction (4.11) is endothermic. If we use $_3^6\text{Li}$ only, the power balance in the reactor is increased, and the reactor can operate with higher efficiency. Using $_3^6\text{Li}$ only, however, incurs the cost of concentrating the $_3^6\text{Li}$, and the use of Li as a natural resource is lowered. From the point of view of generation, use of $_3^6\text{Li}$ only is not economical. Here 'economical' means that the usage and control of neutrons in the reactor system is efficient; neutrons breed with a breeding ratio of greater than unity. One D–T reaction produces one neutron, and by eq. (4.10), one neutron produces one atom of T. Using reaction (4.10) only, the number of neutrons is insufficient, because some of them are absorbed in materials other than $_3^6\text{Li}$, and some are lost through leakage from the reactor. By reaction (4.11), one neutron produces one T atom and one neutron, compensating for the decrease in the neutrons in the reactor. If the number of neutrons is still insufficient, even when $_3^7\text{Li}$ is used in the blanket, Be is mixed with Li. The Be increases the number of neutrons by reaction (4.12). In this case, Be is also part of the fuel for the D–T reaction, because the Be is consumed in the blanket (Fig. 4.10).

### The tritium-breeder and coolant

In order to breed T, most of the blanket consists of Li. There are three ways of breeding T:

(i) metallic Li itself is used;
(ii) molten salts of Li are used;
(iii) solid compounds of Li are used.

Li is liquid at temperatures in the range 186–1380 °C. It is a good carrier of thermal energy because it has high thermal conductivity, high thermal capacitance, and low viscosity. Liquid Li can be used as a coolant, circulating around the blanket and the heat exchanger, as well as the T-breeder.

The superconducting magnet generates a strong magnetic field in the reactor. When a fluid with high electrical conductivity flows across the magnetic field lines, the Lorentz force acts to oppose the flow. Considerable power from the circulating pump is required to make the liquid Li flow against the Lorentz force. This problem can be solved if we use molten salts of Li, which have low electrical conductivities. Chemically, Li is a very active metal and produces many compounds. Of these, Li haloids, especially LiF, are preferable from the thermal and chemical points of view. The breeding ratio of LiF to T is lower than that of pure Li and hence $BeF_2$ is mixed with LiF. The Be in $BeF_2$ increases the number of neutrons in the

**Fig. 4.10.** The first wall and blanket for the 'STARFIRE'.

(a) First wall and blanket     (b) Reactor cross-section

(c) Blanket in detail

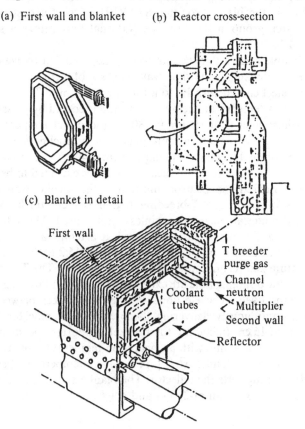

First wall

T breeder
purge gas

Channel

Coolant
tubes

neutron

Multiplier

Second wall

Reflector

blanket, as indicated by eq. (4.12). A mixture of LiF and $BeF_2$ with a molecular ratio of 2:1 is frequently used in the conceptual design of the D–T reactors; this is called the flibe.

There are proposals to use solid compounds of Li such as $Li_2O$, $LiAlO_2$ and LiAl, as the breeder. Cubes of these compounds are placed in the blanket and He gas is circulated through the gaps between them. The Li compounds play the role of T-breeder and He the role of coolant for the blanket. He has zero electrical conductivity so the gas circulates well in the blanket. He also has a small collision cross-section with neutrons, so the gas does not absorb neutron energy directly. When these solid compounds of Li are used in the blanket, the amount of T in the reactor can be reduced. The blanket in such T-breeders will have least effect on other parts of the reactor in the event of an accident. From the point of the safety of the reactor, the amount of T is preferably kept small.

As an example of a blanket which uses solid compounds for breeding T, the blanket for the 'STARFIRE' reactor designed by the research group at the Argonne National Laboratory is shown in Fig. 4.10.

The blanket has a thickness of 68 cm and is cooled by water. Of this 68 cm thickness, the first wall consists of 1 cm of austenitic stainless steel coated with Be to a thickness of 1 mm. The first wall operates at a temperature below 423 °C and is cooled by pressurised water whose inlet temperature is 280 °C and outlet temperature is 320 °C. Next, the neutron multiplier has a thickness of 5 cm, consisting of Be. Its maximum temperature is 490 °C and total mass is 5180 kg. The neutron multiplication factor is expected to be more than 1.2. Behind the neutron multiplier, the second wall has a thickness of 1 cm. The T-breeding region is 4 cm thick. As the structural material, austenitic stainless steel is used. The T-breeder $LiAlO_2$ of grain size of $10^{-7}$ m is used at a temperature between 800–1500 °C, cooled by pressurised water of 15.2 MPa with inlet temperature 280 °C and outlet temperature 320 °C. The T-breeding ratio and the doubling time are expected to be 1.02–1.06 and 2.0–2.3 years, respectively. The He flows through the spaces between the breeder, continuously delivering the bred T. The reflector has a thickness of 15 cm (including the blanket supporter and manifold) and is made of graphite with an operation temperature of less than 800 °C. Austenitic stainless steel (operated at a temperature between 300–400 °C) supports the reflector. The blankets are separated into modules, each of width of 2–3 m and height of 1–3 m.

### 4.1.5. The superconducting magnet

The superconducting magnet is located outside the reactor, serving to build up the magnetic field and so confine the plasma in the reactor centre. The electric power supplied to the magnetic coil is calculated to be 2400 MW, if a copper coil is used to apply the magnetic field in a practical fusion power reactor of 2000 MW electrical output. For the superconducting magnet, a cryogenic machine is required whose electrical power input draws off only a few per cent of the output power of the reactor, and so does not disturb the power balance of the reactor too greatly.

#### Coils for high-energy and high-intensity magnetic fields

The superconducting magnet is characterised by the intensity of the magnetic field produced by it and the energy stored in it. Figure 4.11 plots these values for the superconducting coils already designed and constructed for Tokomak reactors. If we compare them we can see that the superconducting coils for fusion power reactors are much larger than the others.

**Fig. 4.11.** Comparison of superconducting magnets.

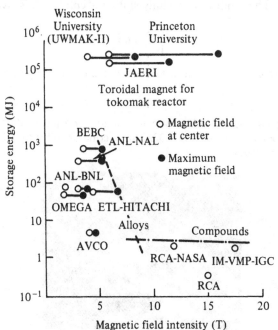

A superconductor is unstable. A small increase in temperature caused by a disturbance results in a decrease in the critical current intensity, which then allows penetration of the magnetic flux in the conductor, inducing an energy loss and further increase in temperature. This cycle produces the instability and eventually leads to loss of the superconductivity. During this process, rapid penetration of the magnetic flux in the conductor is observed; this is called the flux jump. To prevent flux jump, the superconducting coil has many fine superconducting wires of radius of 10–50 μm, implanted in metallic Cu or Al. An example of this kind of superconductor is shown in Fig. 4.12. The base metal Cu or Al is called the stabiliser. When the superconducting wires lose their superconductivity, the highly conductive Cu or Al forms a parallel circuit and reduces the abnormal heat generation, thus causing the cooling by the coolant to be more effective.

The superconducting magnet also suffers from mechanical disturbance due to electromagnetic force.

The large size of the superconducting magnet requires the use of a fully-stabilised conductor, one which prevents the temperature from increasing above the critical value, at which all the electric currents in the superconducting wires change so as to flow in the stabiliser. The condition for a fully-stabilised conductor is given by

$$\frac{\rho}{A} I^2 < qS, \qquad (4.13)$$

where $I$ (in units of A) is the maximum current in the superconductor,

**Fig. 4.12.** Cross-section of a fine, multi-line superconductor. 1.2 × 4.0 mm; NbTi, 17 × 3025 lines; critical current (in liquid He); 1050 A.

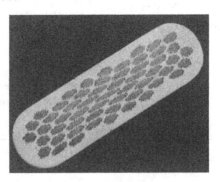

$\rho$ ($\Omega$ m) is the specific resistivity of the stabiliser, $A$ (m$^2$) is the cross-sectional area of the stabiliser, $q$ (W m$^{-2}$) is the thermal energy conveyed by the coolant, and $S$ (m) is the surface area of cooling per unit length. Equation (4.13) indicates that the thermal energy conveyed by the coolant from the conductor exceeds the Joule heating in the fully-stabilised conductor. In the case of a fully-stabilised conductor, the superconductor recovers its super-conductivity and continues to operate safely, once the cause of the disturbance is removed.

Table 4.3 analyses the materials used in superconductors that are in current practical operation. Fully-stabilised conductors have been developed for the alloys listed in the table, and the chemical compounds are being investigated.

The Japanese Atomic Energy Research Institute has constructed a superconducting coil TMC-1 (Test Module Coil 1) (see Fig. 4.13)

Table 4.3 *Practical superconducting materials*

| Material | | Critical temperature | Critical intensity of magnetic field (4.2 K) |
|---|---|---|---|
| Alloy | NbZr | 9–11 K | 7–9 T |
| | NbTi, NbTiZr, NbTiTa | 8–10 K | 9–12 T |
| Compounds | V$_3$Ga | 14.5 K | 21–23 T |
| | NbSn | 18.3 K | 22–26 T |

**Fig. 4.13.** Test of the TMC-1.

for a high-intensity magnetic field. The circular coil of TMC-1, whose inner radius is 60 cm and outer radius is 185.4 cm, is made of $Nb_2Sn$ and NbTi, and is cooled in a pool of liquid He at 4.2 K. In the magnetic field of 4.3 T induced by the CTC (Cluster Test Coil) and CBC (Cluster Break-up Coil), an electric current of 6 kA flows in TMC-1 coil and generates a magnetic field whose intensity is 11.1 T. Although one turn of the inner-most coils is heated and loses superconductivity, this superconductivity is restored after 6.3 s. This restoration of the superconductivity shows that $Nb_2Sn$ has a high critical value for the intense magnetic field.

### The structure and support of the electromagnetic force

If we review the discussion in chapter 2 on the configuration of magnetic fields to confine the plasma it can easily be seen that the helical coil for the stellarator and the baseball coil for the mirror machine, for example, have extremely complex shapes. To construct such a complex coil is not easy, and to support it against the electromagnetic force that acts on it is also difficult. For a fusion power reactor, various devices, such as heating facilities, fuel injectors, evacuation systems, and the manifolds for the coolants, must be located in the small gaps among the coils. The space available for the structural supporter against the electromagnetic force is thus limited. Nevertheless, the supporting system must be designed to cope with all possible accidents, including the case in which the electromagnetic forces are out of balance due to changes in the current in the superconducting coil.

### Pulse operation

The transformer coil for the Tokomak and the compression (one turn) coil for the theta-pinch machine are pulse-operated. The pulsed magnetic fields overlap, whether due to disturbance of the plasma or to leaking fields of other coils, onto the fields produced by the toroidal coil of the Tokomak, by the helical coil of the stellarator, and by the baseball coil of the mirror machine, although the aim is to operate them steadily. The superconductor for the pulsed power must reduce the energy loss of the magnetic hysteresis; the required condition is more severe than that for a fully-stabilised conductor to operate against the flux jump of the DC current.

### Radiation damage

The superconducting magnet for a fusion power reactor is irradiated by the neutron flux and the $\gamma$-rays that leak out from the shield layer. The coil can suffer from the following problems as a result of this irradiation:

   (i) loss of the superconductivity, reduction of the super-conductivity of the materials used in the superconductive wire, reduction of the electrical conductivity of the stabiliser, and reduction of the insulating ability of the insulator;

  (ii) heat generation in the coil.

What conditions result in the phenomena described above depend on the intensity of the irradiation. In other words, the phenomena depend on the thicknesses of the blanket and the shield layer. If the thickness of the shield layer increases, the intensity of the irradiation decreases, but the dimensions of the reactor become large and the construction cost high. If the thickness of the shield layer decreases, the lifetime of the superconducting magnet is shortened and the input power required for the cryogenic machine increases. A design for the reactor which optimises these conflicting considerations is required.

    The neutron flux which irradiates the superconducting magnet is very low in comparison with that for the first wall; it is of the order of $10^9 \, \text{n cm}^{-2} \, \text{s}^{-1}$. At present, irradiation tests on the super-conductor in the fusion reactor are being carried out, but systematic results have not been obtained yet. The loss of superconductivity due to irradiation under a cryogenic low temperature is restored once the temperature of the coil increases to the normal value. This phenomenon is called the annealing effect. However, the giant superconducting magnet for a fusion power reactor cannot frequently increase its temperature to this normal value. Moreover, the irradiation test has been performed on a superconductor with a short length, which is quite different from the situation which would obtain given the size of a real reactor.

### Remote handling

    The structural supporter of the superconducting magnet needs sufficient mechanical strength to cope with huge electromagnetic forces. Thus the structural materials of the fusion reactor will occupy a large space and be very heavy. On the other hand, the materials used in the reactor, especially the first wall, will face severe damage from sputtering, blistering and swelling. Thus the first wall has to be replaced rather frequently. Unfortunately, radio-activity will have been induced in these materials, so they cannot be approached directly. To repair the reactor, remote handling techniques must be developed. A reactor using magnetic confinement will have a complicated structure with complex manifolds and various magnetic coils. The remote handling of such a reactor needed to replace its parts will present great difficulties.

## 4.2. Design of fusion power plant

Examples of power plant for nuclear fusion reactors using magnetically-confined plasma are presented in this section. How the technical problems described in the preceding section are being solved are also described.

The examples are chosen for a torus type, the Tokomak, on which research has advanced furthest, and for an open-end type, with the mirror type as a representative case.

### 4.2.1. Design of fusion power plant of the Tokomak type
#### An example of the design

In the Tokomak, the scaling laws which give the relationships between the various parameters of the device and the confinement measure (density multiplied by the confinement time $n\tau$) have been established, and the parameters of the power plant can be estimated by extrapolation from these scaling laws. The scaling laws are as follows:

(i) theoretical,

$$n\tau \propto n^{\frac{1}{3}}a^4 B^5 As^{\frac{3}{2}}T^{-\frac{7}{2}};$$

(ii) empirical,

$$n\tau \propto n^{\frac{1}{3}}a^{\frac{3}{2}}B.$$

Here $T$, $a$, $As$, and $B$ denote the temperature, the minor radius, the aspect ratio (major radius/minor radius) of the plasma, and the magnetic flux. Scaling law (i) takes the trapped-particle instability into consideration and is valid for plasma which is either collisionless or has few collisions. Scaling law (ii) is obtained from the results of experiments.

Present research on the Tokomak seeks to verify the scaling laws and to realise plasma which will satisfy the Lawson criterion by scaling-up of the facility (see Fig. 2.30). The fusion power plant is being designed conceptually on the basis of developments in this research.

The conceptual structure of the Tokomak-type power plant is shown schematically in Fig. 4.14. Although the Tokomak itself is fundamentally the same as that depicted in Fig. 2.14, the power plant includes the elements shown in Fig. 4.1 which are required for the plant. Commonly, the iron yoke for the torus is omitted, and the torus for the reactor has its major axis in a vacuum. The transformer coils induce the electric current in the plasma. The coils to control the position of the plasma determine this position via the force given by the product $B_\perp \times I_p$ of the magnetic field $B_\perp$ produced by the coils and the plasma current $I_p$. All these coils consist of superconducting magnets. It is required that the transformer coils and the coils to control the plasma position be pulse-operated.

Table 4.4 gives concrete examples of the design for the Tokomak-type power plant. From the loadings of the first wall, the materials used in the T-breeder, the cooling methods, the intensities of the magnetic flux, and the choice of materials for the superconductor indicated in this table, the considerations and aims of the designers can be appreciated. Several important design points will now be noted.

### The power density and wall loading

Let us investigate here the relation between the power density and the wall loading. The power density $P_d$, which is the fusion output energy released per unit time and unit volume of the reactor, is given by eq. (1.33):

$$P_d = \tfrac{1}{4}n^2 \langle \sigma v \rangle E_f, \tag{4.14}$$

where $\langle \sigma v \rangle$ is a function of the ion temperature $T_i$. If $T_i$ is assumed to be constant, eq. (4.14) becomes $P_d \propto n^2$, which shows that $P_d$ is high when $n$ is high. The $\beta$-value, introduced in section 2.1.6 as the

**Fig. 4.14.** Conceptual structure of the Tokomak-type reactor.

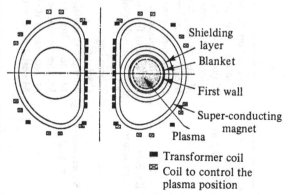

Table 4.4 *Examples of designs for Tokomak-type power plant*

| | | JAERI* | Univ. of Wisconsin | | Princeton Univ. |
|---|---|---|---|---|---|
| | | | UWMAK-II | UWMAK-III | |
| Output | Thermal output power (MW) | 2000 | 5000 | 5000 | 5305 |
| Power | Electric output power (MW) | 800 | 1700 | 1985 | 2030 |
| Central plasma | Major radius (m) | 10 | 13 | 8.1 | 10.5 |
| | Minor radius (m) | 2 | 5 | 2.7 | 3.2 |
| | Mean ion temperature (keV) | 15 | 15.2 | 19–23 | 30 |
| | Mean ion number density (particles/m$^3$) | $1.0 \times 10^{20}$ | $6.46 \times 10^{19}$ | | $5 \times 10^{19}$ |
| | Confinement time (s) | 1.8 | 4.0 | 1.6 | 3.83 |
| | Plasma current (MA) | 8 | 14.9 | 15.8 | 14.6 |
| Reactor structure | Material in first wall | Mo alloy | 316SS | TZM** | PE-16*** |
| | Neutron wall loading (MW/m$^3$) | 1.4 | 1.16 | 2.5 | 1.76 |
| | T-breeder | Li$_2$O | Li$_2$Al$_2$O$_4$ | Li | Flibe |
| | T-breeding ratio | 1.3 | 1.19 | 1.25 | 1.04 |
| | Coolant | He | He | Li, He | He |
| Superconducting magnet | Number of divisions in the coil | 24 | 24 | 18 | 48 |
| | Magnetic flux at centre (T) | 6 | 3.57 | 4.05 | 6 |
| | Maximum magnetic flux (T) | 11.5 | 8.8 | 8.75 | 16.0 |
| | Stored energy (GJ) | 160 | 238 | 108 | 250 |
| | Material of superconductor | Nb$_3$Sn–NbTi (hybrid) | NbTi | NbTi | Nb$_3$Sn |

* Japan Atomic Energy Research Institute
** TZM (Mo: 99.4 %, Ti: 0.5 %, Zr: 0.08 %)
*** PE-16 (Ni: 43%, Fe: 39 %, Cr: 18 %)

utilisation factor of the magnetic field, can be described as

$$\beta = \frac{nk(T_i + T_e)}{B^2/2\mu}.$$  (4.15)

If $T_i$ is constant, eqs (4.14) and (4.15) give

$$P_d \propto \beta^2 B^1.$$  (4.16)

Equation (4.16) explains the economy of the fusion power plant. The power density $P_d$ becomes high when $\beta$ is high if $B$ is constant, and $P_d$ becomes high when $B$ is large if $\beta$ is constant. When $P_d$ is high, the reactor can be small, provided that the output power is constant. On the other hand, the wall loading $P_w$ becomes large when $P_d$ is high, provided that the output power is constant. In the early stage of the conceptual design of the fusion power reactor, a high wall loading $P_w$ such as $10\ \mathrm{MW/m^2}$ was chosen. For such a high wall loading, the lifetime of the first wall is extremely short. Recently, more moderate values such as $1-3\ \mathrm{MW/m^2}$ have been chosen for $P_w$, as shown in Table 4.4. Thus Table 4.4 shows that UWMAK-III and UWMAK-II have the same thermal output power, while the ratio of volumes is $1/5.5$, and the wall loading of UWMAK-III is about twice that of UWMAK-II. As is clear from these examples, the choice of values for $P_d$ and $P_w$ is one of the most important problems, one that is related to the choice of wall materials in the general design of the reactor.

Cooling the reactor was one of the severe technical difficulties arising from a high wall loading such as $P_w = 10\ \mathrm{MW/m^2}$. For $P_w = 1-3\ \mathrm{MW/m^2}$, however, cooling the reactor is not so difficult. Thus the freedom in the choice of the blanket material, the coolant material, and the reactor structure is enlarged. The main points of emphasis in recent designs for the reactor are that

(i) the safety of the reactor is increased by decreasing the amount of tritium in the blanket;
(ii) the reactor has the minimum after-effects from accident damage, so restoring it to operation is easy and fast.

### The toroidal magnet

The general description of the superconducting magnet for the fusion power plant was given in section 4.1.5. Here, several specific problems in the design of the toroidal magnet will be investigated.

First to consider is the reason why the magnet has a D-shape. The toroidal coil is divided into 24–28 pieces and is located in a circle surrounding the torus plasma. From Ampère's law, the generated magnetic field $B$ is given by

$$B = \frac{\mu AT}{2\pi r}, \qquad (4.17)$$

which shows that $B$ is inversely proportional to the minor radius $r$. Here $\mu$ is the magnetic permeability in a vacuum and $AT$ is the total number of ampere turns of the coil. Figure 4.15(a) shows the circular coil. When electric current flows in a single circular coil, the expanding force acts on the coil outward in the radial direction. In the case where the magnet forms a torus, the expanding force in the radial direction has a different value depending on the azimuthal angle along the circle of a minor radius; this induces the bending stress in the toroidal coil. The tension on the coil can be supported by the superconducting material itself but support for the bending stress must be provided with care. In view of this, there is a proposal to use a coil shape which receives no bending stress. No bending stress acts on the coil when its curvature is proportional to $r$, as shown in Fig. 4.15(b). The shape given in Fig. 4.15(b) is called the pure-tension D-shape. Although the pure-tension D-shape can be obtained by analysis, the shape can also be formed experimentally if an electric current flows along the sub-axis of a torus in a deformable coil supported by two points at A and B in Fig. 4.15(b). The pure-tension D-shape is clearly appropriate for the toroidal coil, but the coil must be designed to take into account the bending stress that is generated by disturbance of the magnetic field due to an accident. In actual designs, a modified D-shape is frequently used. Such a modified D-shape has a small bending stress, but the shape is preferred in practice because it takes into consideration the question of compatibility with the arrangement of other structural parts.

**Fig. 4.15.** Various shapes of toroidal magnets.

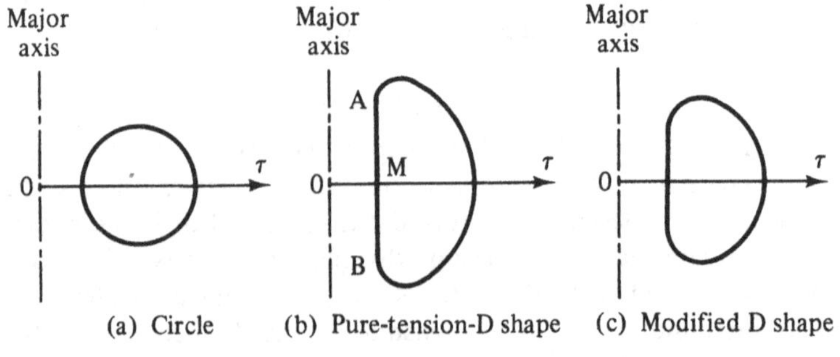

(a) Circle        (b) Pure-tension-D shape        (c) Modified D shape

For the value of the magnetic field at the centre, 3–6 T is chosen as Table 4.4 indicates. Such a magnetic field intensity is less than the critical field intensity of the alloys in Table 4.3. In the design of the superconducting magnet, the material for the superconductor is selected not on the basis of the value of the magnetic field at the centre (at 0 in Fig. 4.15(b)), but rather the empirical maximum value (at M). The magnetic flux at the centre, the maximum magnetic flux and the material for the superconductor given in Table 4.4 account for this as described above.

*Steps toward the development of the reactor*

The design parameters tabulated in Table 4.4 are for practical fusion power reactors. But such practical reactors cannot be constructed at present. To turn the various experimental facilities for attaining $Q \approx 1$ (JT-60 (Japan), JET (EURATOM), TFTR (USA) and T-15 (USSR); see Table 2.2 and Fig. 2.30), into practical reactors, by closing the current physical and technical gaps, developmental research must advance several steps in succession. Typically, these steps are:

(i) *Test facilities for engineering reactor plasma:* needed to produce plasma with a high-$Q$ value, through which self-ignition is possible, and to perfect the technology for controlling the plasma.

(ii) *Test reactors:* needed to produce steady power from a facility which has the same main components as a reactor, and to test the functioning and to ensure the safety of these components.

(iii) *Prototype reactors:* needed to perfect all the components of the electric power plant taken together, and to test the power plant as a whole and to establish its safety.

(iv) *Proving reactors:* needed to test the overall ability of the electric power supply system from the points of view of social relations, safety, reliability, and economy.

The steps described above are the projected ones. In reality the steps will be taken according to the processes involved and the developmental results obtained. As described in chapter 2, with present Tokomaks designed to attain $Q \approx 1$ there are theoretical and technological difficulties that stand in the way of converting them into practical reactors. Attaining $Q \approx 1$ means that the input beam and wave energies of the Tokomak plasma are roughly equal to the output fusion thermal energy. The ratio between the input electrical energy of the Tokomak and its fusion output electrical energy is currently less than 0.01. This ratio must be greater than 100 in a practical Tokomak, if the large amount of energy consumed in the construction of the giant Tokomak facility is taken into account. Thus a practical Tokomak must be scaled up enormously, in accord with its scaling law. To heat the plasma in the large Tokomak, a great amount of energy must be supplied, which disturbs the stability of the plasma. A large Tokomak must be operated steadily, because the energy consumed in starting up is enormous. But present Tokomaks suffer from many instabilities which hinder steady operation, and have as yet no means of sustaining the toroidal current continuously. As this chapter has shown, the plasma will cause severe damage to the first wall as a result of sputtering, blistering, swelling and induced radio-activity, even with the moderate wall loading of 1–3 MW/m$^2$. Tests on a wall subjected to strong particle and radiation fluxes have not yet been carried out and no material has yet been proved to be capable of serving in the first wall to enable the wall to play its role with a meaningfully long lifetime.

Whether the present Tokomaks can be turned into practical fusion power reactors will be decided only in the distant future.

### 4.2.2. Design of mirror-type fusion power plant

The mirror-type fusion power reactor has the advantage of being operated steadily. However, in the mirror field, the plasma flows out from the open ends of the magnetic field, and it is difficult for the reactor to attain power balance. Thus some preventive measures must be taken to control energy loss from the open ends. The main methods are:

(i) electromagnetic fields are applied to the open ends to suppress the outflow of plasma;

(ii) using the method of MHD direct energy conversion, the energy of the plasma flowing out the open ends is returned with high efficiency;

(iii) the mirror fields are connected with each other to form a straight line or a torus, and the energy losses from the open ends decrease.

Here we shall give an example of the conceptual design of the mirror-type power reactor which uses the method of direct energy conversion at the open ends. Figure 4.16 shows its conceptual structure (except for the direct energy conversion stage). The plasma is confined in the region of the minimum magnetic field by the Yin–Yang coil, as in Fig. 4.16(a). The Yin–Yang coil is made of superconducting magnets, inside which the blanket is set along the magnetic field lines, as in Fig. 4.16(b). Liquid Li flows in the blanket to play the roles of T-breeder, coolant, and neutron attenuator. In a reactor with open ends, the liquid Li can flow along the magnetic field lines, and the energy loss induced by the electrically-conductive liquid crossing the magnetic field lines can be decreased below the maximum allowable level.

The flow of power in the mirror-type reactor is shown in Fig. 4.17, and the main parameters of the reactor are given in Table 4.5. High-energy fuel must be injected into the mirror to compensate for the energy loss from the open ends; hence large neutral-beam injectors must be devised. As a result, the $Q$ value of the reactor becomes low. In Table 4.5, $Q = 1.2$ is chosen. The energy-multiplying constant $m$ in the blanket is determined by the fact that the thermal energy is released in the ${}_{3}^{6}$Li when neutrons react with these atoms to produce tritium. The numbers in the brackets in Fig. 4.17 show the values for $m = 2.0$ instead of $m = 1.1$. Since the output electrical power of the reactor for $m = 2.0$ is twice that for $m = 1.1$, the energy-multiplying constant in the blanket plays an important role in determining the output power of the mirror-type reactor.

According to the structure of the blanket and the ratio of the composites of ${}_{3}^{6}$Li and ${}_{3}^{7}$Li, the value of $m$ can be changed a little. When the blanket includes the fission materials (such a reactor is called the fusion–fission hybrid reactor, and the reaction that occurs

**Fig. 4.16.** Structure of the mirror-type fusion power reactor.

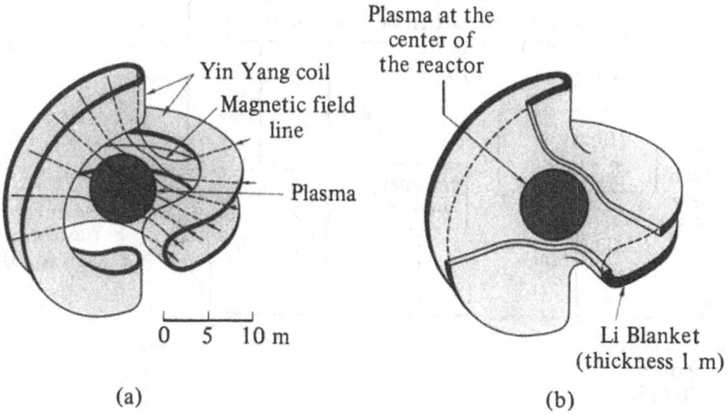

Plasma at the center of the reactor

Yin Yang coil

Magnetic field line

Plasma

0   5   10 m

(a)

Li Blanket (thickness 1 m)

(b)

in the blanket is

$$^2_1H + ^6_3Li + 3\,^{238}_{92}U = 2\,^4_2He + 2\,^{239}_{94}Pu \ (400 \ MeV)$$
$$+ \,^{90}_{40}Zr + \,^{145}_{60}Nd + 12e + \,^1_0n + 224 \ MeV),$$

$m \sim 30$ can be achieved and the power balance of the reactor is raised. The hybrid reactor not only enhances the power balance but also produces the fissionable fuel Pu. But of course the hybrid reactor will also produce the strong radio-activities of the fission products such as Zr and Nd.

The mirror-type reactor will be operated at a high temperature, because fewer ions fall into the region of loss-cone, as a result of fewer collisions in the high-temperature plasma. The average temperature of the plasma is expected to be 600 keV, and the $\beta$-value for the mirror-type reactor is also expected to be high.

The neutron output power of 470 MW, as the product of the fusion reaction, is converted to electrical power from thermal power at a conversion rate of 45 %. The ion power of 610 MW which flows

**Fig. 4.17.** The flow of power in the mirror-type fusion electrical power plant. (The numbers are in MW; those in brackets are for $m = 2.0$; 80 % of the fusion energy is given off via neutrons and 20 % via $\alpha$-particles.)

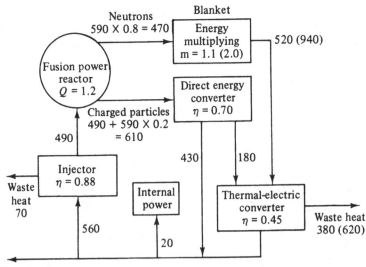

out the open ends of the mirror field is collected and changed to electrical power by the MHD direct generator, at a conversion rate of 70 %. The thermal energy generated in the process of direct energy conversion of the ion power is recovered by thermal–electrical energy conversion.

The electric power generator using the method of direct energy conversion consists of three parts as is shown in Fig. 4.18.

Table 4.5 *An example of a design for a mirror-type fusion power reactor*

| | |
|---|---|
| Output power | |
|   Fusion output | 590 MW |
|   Neutron output | 470 MW |
|   Ion output | 610 MW |
|   Pure electric output | 170 MW |
|   Injector output | 490 MW |
| Plasma | |
|   $Q$ value | 1.2 |
|   Volume | 130 m$^3$ |
|   Density | $1.2 \times 10^{20}$ (particles/m$^3$) |
|   $m$-value | 0.45 |
| Others | |
|   Magnetic field at the centre* | 5 T |
|   Mirror field* | 15 T |
|   Wall loading | 1.7 MW/m$^2$ |
|   Average injection energy | 550 keV |

* Values in vacuum: mirror ratio in vacuum = 3, mirror ratio with plasma = 7.7

**Fig. 4.18.** A direct electrical power generator for the mirror-type fusion power reactor.

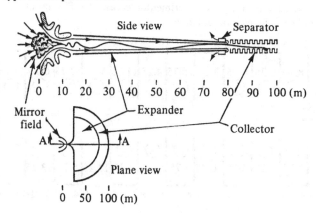

(i) *The expander:* The magnetic field lines expand from the reactor end in the fan-shape expander. The intensity of the magnetic field decreases outward. The intensity of the magnetic field at the reactor end is 10 T, but at the end of the expander it is of the order of 0.01 T. The plasma flowing out from the reactor end initially has kinetic energy $W_\perp$ perpendicular to the magnetic field. The kinetic energy of the plasma particles decreases with the expansion of the expander, as explained in section 2.1.1. (Equation (2.13) shows that $W_\perp/B = M_m = $ const.) The number density of the plasma decreases from $10^{20}$ particles/m$^3$ to $10^{11}$ particles/m$^3$ following the expansion. With this expansion, the plasma loses the properties of a continuous medium and behaves as a group of particles. The radius of the fan-shape expander needs to be about 80 m to achieve its purpose.

(ii) *The separator:* The separator is located at the outer edge of the expander. At the separator, the magnetic field lines, which guide the charged particles along the lines from the reactor exit to the separator, are abruptly bent perpendicularly. Electrons flow out along the magnetic field lines from the separator, but ions, which have greater inertias, cannot continue to flow along the best field lines, and so fall into the collector, which collects the kinetic energy of ions as electrical power. In the expander, electrons and ions have the same mean velocity. Thus the kinetic energy of an electron is negligible in comparison with that of an ion.

(iii) *The collector:* The collector is an array of electrodes whose voltages increase at their outer side. The collector which a research group at the Lawrence Livermore National Laboratory has used in its experiments is shown in Fig. 4.19. Among the collecting electrodes with the ⊓ shape, rod electrodes with high voltage (full circle) and

**Fig. 4.19.** Part of an experimental collector. (The numbers indicate the electric potentials.)

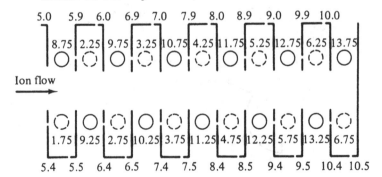

Table 4.6 *Comparison of three alternative cycles*

| Parameter | Saturated steam cycle | Simple SH steam cycle | SH + RH steam cycle |
|---|---|---|---|
| Fusion power (MW) | 479 | 479 | 479 |
| Maximum steam temperature (°F) | 1000 | 1000 | 1000 |
| Generator power (MW) | 169 | 206 | 229 |
| Direct converter power (MW) | 39 | 39 | 39 |
| Recycle power (MW) | 126 | 113 | 92 |
| Net power (MW) | 81 | 131 | 176 |
| Reject power to cooling tower (MW) | 509 | 478 | 451 |
| Power multiplier factor, $Q_e$ | 1.64 | 2.16 | 2.92 |
| Steam consumption rate (lb/kWh) | 9.59 | 7.12 | 5.80 |
| Steam heat rate (BTU/kWh) | 10840 | 9240 | 8520 |
| Steam cycle efficiency (%) | 31.5 | 37.0 | 40.0 |
| Plant efficiency (%) | 13.8 | 21.6 | 28.1 |

rod electrodes with low voltage (dotted circle) are arranged alternately. The average voltage of the rod electrodes becomes higher outside. The velocity of an ion which falls into the collector is decreased by this increasing voltage, and the ion halts in a potential well. The ion eventually deposits its charge on a rod electrode, maintaining a high direct voltage. To increase the energy conversion rate, an ion must be trapped on an electrode whose potential is nearly equal to the original kinetic energy of the ion.

There are proposals in which the magnetic field expands in two directions, and a Venetian-blind-type grid is inserted to suppress the charge effect of particles in the resultant space.

In the mirror-type fusion power reactor, the $Q$ value, the energy conversion rate of the direct power generator, and the efficiency of the beam injector, have a large effect on the total power balance, as indicated in Fig. 4.17. Further investigation of the method of direct energy conversion needs to be carried out. How much is the efficiency of the reactor improved? In the Lawrence Livermore National Laboratory a power supply system using steam cycles has been studied. Table 4.6 shows the power balance for three cycles, the saturated steam cycle, the simple superheated steam cycle, and the superheated and reheated steam cycle.

The mirror-type reactor offers the possibility of advantages such as steady operation, low wall loading and so on, in comparison with the Tokomak-type reactor. However, the plasma parameters so far achieved in the mirror field are much inferior to that in the Tokomak. It seems that it will be only in the distant future that a mirror field can be used in a practical fusion reactor.

## 4.3. Power reactors using inertial-confinement fusion

Descriptions of the reactor for inertial-confinement fusion are given in this section. The examples are provided by reactors using a laser and a light-ion beam (LIB) as drivers.

### 4.3.1. Design of laser fusion power reactors
*Characteristics of the power reactor for inertial-confinement fusion*

The structure of the fusion power reactor which uses plasma confined by inertia is simplified. Because there is no superconducting magnet, the shape can be spherical, or whatever is preferred. Much freedom can be exercised in designing the structure, which can be smaller by comparison with that for magnetic-confinement fusion. Since the fusion reaction occurs over a short period, impurities in the fuel do not cause major difficulties stemming from radiation loss. On the other hand, the repeated impulses caused by micro-explosions of the pellets has serious consequences for the first wall, the structure of the blanket, and protection of the parts of driver. New technologies are required for pellet injection into the reactor cavity, the control of the pellet path in the reactor cavity, and the focusing of many beams onto the surface of the pellets.

**Fig. 4.20.** The temperature increase of the reactor surface of the first wall in inertial-confinement fusion.

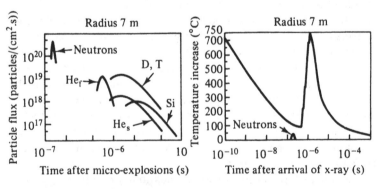

(a) Particle fluxes injecting       (b) Temperature increase of
    on the wall surface             the reactor surface

The calculated data for the increases in the temperature of the surface of the wall generated by micro-explosions of the pellets are as follows. Of the energy flux onto the wall, 77 % is from neutrons, 1 % from X-rays and 22 % from the atoms and ions, for instance, unburnt D and T, He as a fusion product, and pellet materials surrounding the fuel. For the micro-explosion of a pellet whose fusion output is 100 MJ in a spherical reactor cavity of radius 7 m, the energy fluxes impinging on the wall are as shown in Fig. 4.20. The use of Si as the material of the pellet which covers the fuel is illustrated in Fig. 4.20(a). It is clear that the temperature of the wall increases almost instantaneously. Melting and evaporation of the wall material will thus occur at the same instant. For this increase in wall temperature, X-rays and charged particles are responsible, rather than neutrons. Accordingly it is desirable that these X-rays and charged particles do not hit the wall directly. The following proposals have been made to protect the wall:

(i) The charged particles are guided out of the reactor along the applied magnetic fields.

(ii) Make the first wall of a porous medium and liquid Li, which flows in at the back and leaks out at the front, to make a thin liquid film of 1–2 mm thickness. This film will protect the first wall from damage caused by evaporation resulting from the irradiation of the fluxes. This type of wall is called a wet wall. The repetition rate of the reactor is limited by the time it takes for the vapour condensates to form the liquid Li layer on the surface of the wall; this rate is of the order of 0.1 Hz.

(iii) The front of the first wall is covered with thick liquid Li or a molten salt of Li (e.g. flibe). This kind of wall is called the waterfall type-1. It has been suggested that the micro-explosion of the pellets occurs in a hole of the eddy of the liquid Li flow. It takes 1–10 s for the flow of the waterfall to be restored after the perturbations caused in it by the micro-explosion of the pellet.

(iv) Circulation of a heavy gas (such as Ne or Xe) of 0.5 Torr in the reactor cavity would suppress direct impact of the charged particles and of soft X-rays on the surface of the wall. The energy absorbed in the gas can be re-emitted with a short time lag.

The structure of the reactor depends on which of these methods the designer chooses.

*The solase-type laser fusion reactor*

The design of the laser fusion reactor has been carried out by a research group at the University of Wisconsin using method (iv) above; the reactor is called the solase type (see Fig. 4.21). The parameters chosen in the design are listed in Table 4.7 along with those for the waterfall-type reactor designed by a research group at the Lawrence Livermore Laboratory.

The reactor cavity is a sphere of radius 6 m. The first wall and the structural material of the blanket are made of a graphite compound; the outer side is supported by Al. These materials have low radio-activities, as will be discussed in section 4.4.2, and are chosen with the safety of the reactor in mind. The reactor itself has the shape of a simple sphere and can be constructed of materials such as graphite compounds and Al, which are not exceptionally strong. Around the back of the first wall, $Li_2O$ is circulated to breed T, and also to cool the reactor. The pellet gain is chosen as low as 150, since the designers consider that a target with high pellet gain is not yet fully proven. Twelve beams are projected at the target to give uniform irradiation. The requirement for the driver efficiency is severe for the solase-type reactor due to the low (but not exceptionally low) pellet gain. The value for the driver efficiency chosen for the design is 6.7 %. (As shown in the example of a waterfall-type reactor designed by the Lawrence Livermore Group,

**Fig. 4.21.** Solase-type reactor for laser fusion.

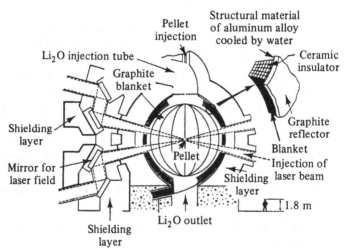

the requirement for the driver efficiency is lowered if the pellet gain is higher.) The circulation of Ne gas at 0.5 Torr in the reactor cavity and in the guide tubes, through which the laser beams pass, protects the first wall and mirrors from damage (see Fig. 4.22). Calculations show that a mirror made of Cu of 1 mm thickness on an aluminium plate can be protected from damage by the Ne gas if the distance from the reactor centre to the mirror is 15 m. The fusion output energy per shot of 150 MJ is so small that wall protection can be ensured with a light construction. However, a high repetition rate is required to obtain practical fusion output power. The solase-type

Table 4.7 *Power reactor for laser fusion*

|  | Solase type (Univ. of Wisconsin) | Waterfall type (Lawrence Livermore Laboratory) |
|---|---|---|
| Thermal power output (MW) | 3340 | 1200 |
| Gross electrical power output (MW) | 1334 | 460 |
| Circulated electrical power (MW) | 334 | 80 |
| Net electrical power output (MW) | 1000 | 380 |
| Total efficiency (%) | 30 | 32 |
| Neutron wall loading (MW/m$^2$) | 5 | |
| Laser energy (MJ) | 1 | 1 |
| Driver efficiency (%) | 6.7 | 2 |
| Number of beams | 12 | 4 |
| Repetition rate (Hz) | 20 | 1.5 |
| Pellet gain | 150 | 700 |
| Fusion thermal output energy per shot (MJ) | 150 | 700 |
| T-breeder | Li$_2$O | Li |

**Fig. 4.22.** Structure for protecting the tube used for injection of laser light.

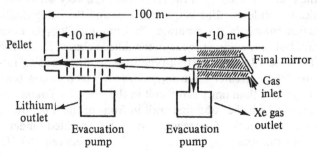

reactor is designed to have a repetition rate of 20 Hz, which may well be possible because the first wall is made of solid material.

For repetitive operation, the laser medium must be a gas. In order to increase the hydrodynamic efficiency of the pellet, it is desirable that the wavelength of the laser light be short. A recent strong candidate for the driver is the KrF (excimer) ultraviolet (UV) laser, whose energy is absorbed classically in the target and is not transferred to high-energy particles but rather to the thermal energy of the electrons. But it is difficult to obtain KrF laser light of high intensity (at most $1 \text{ J/cm}^2$), high conversion rate (the theoretical maximum is 5 %), and short pulse width (more than 100 ns). A pellet gain of more than 100 is also difficult to realise due in general to the position of the energy deposition in the case of lasers. At the Lawrence Livermore Laboratory, Shiva Nova (at present the largest Nd glass laser at 100 kJ and wavelength of 0.3 $\mu$m) gives a fusion output thermal energy of the order of 10 J per shot. The gap between present technology and that required for a practical reactor is thus very large in the case of laser fusion.

### 4.3.2. The waterfall-type reactor for light-ion beam fusion

Recent developments in the pulsed-power technique enabled the Sandia National Laboratories to construct the PBFA-II particle-beam fusion accelerator, which delivers 36 Li beams with a total beam energy of 4 MJ and a pulse width of 20 ns. This machine is almost on the scale of a practical accelerator for a light-ion beam (LIB) reactor. An example of the design for an LIB reactor is shown in Fig. 4.23. The reactor vessel consists of a cylinder which rotates around a horizontal axis. Through the action of centrifugal force, the flibe flows along the wall of the cylinder. In the case of a waterfall-type reactor, it is usually difficult to ensure that the ceiling area of the reactor wall is covered with the liquid. A reactor with a rotating vessel solves this problem through the action of a centrifuge. For the sake of the safety of the reactor (Li is a very active material chemically, while the flibe is quite stable) and the compatibility of the target materials (to separate Pb from $Li_{17}Pb_{83}$ is normally difficult, but in the flibe it is easy), the flibe is chosen as the coolant and T-breeder. In the waterfall-type reactor, the breeding ratio of T in the flibe is greater than unity, although to obtain a breeding ratio of greater than unity is difficult in the case of a Tokomak-type reactor, in which the solid first wall in front of the flibe attenuates the neutron energy. The reactor must be operated under such conditions that the temperature of the flibe is between 550 °C and

750 °C, to ensure sufficient mobility. As for the material of the solid wall of the reactor, Ni alloys, e.g. 'Inconel 713C', are employed for the flibe instead of Vd alloys for the liquid Li. The reactor cavity is filled by Ar gas at a pressure of 0.1 Torr. The fusion energy released from the target (except those conveyed by neutrons) is absorbed in the Ar gas to form a fireball, which gives a time lag in the wall loading.

As Fig. 4.24 shows, the heat is exchanged from the flibe to another molten salt $NaF-BF_3$, and again from $NaF-BF_3$ to the steam, which then drives the turbine and generates electric power.

The cryogenic hollow-shell target of 6 mm radius consists of three layers of Pb (mass $M_{Pb} = 120$ mg, thickness $\delta_{Pb} = 23.4\ \mu m$), Al ($M_{Al} = 184$ mg, $\delta_{Al} = 151\ \mu m$) and D–T fuel ($M_{DT} = 21.5$ mg, $\delta_{DT} = 27.5\ \mu m$) (see Fig. 3.58). The target materials, Pb and Al, have small induced radio-activities.

**Fig. 4.23.** Power reactor for LIB fusion.

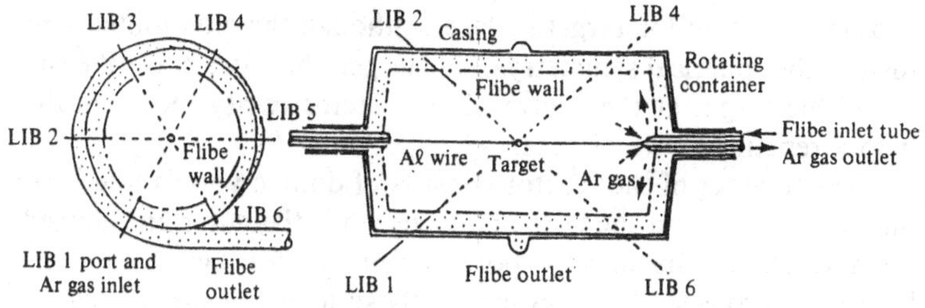

**Fig. 4.24.** A heat exchange system

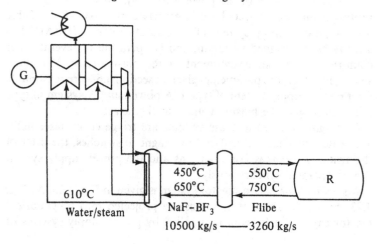

Table 4.8 *Power-supply systems*

| Type | Stored energy | Output voltage | Pulse width | Number of modules |
|---|---|---|---|---|
| 1 | 2.7 MJ | 10 MV | 30 ns | 6 |
| 2 | 1.7 MJ | 5 MV | 30 ns | 6 |
| 3 | 2.6 MJ | −1 MV | 200 ns | 1 |

The six proton beams of a total energy of 7.9 MJ are projected at the target for a period of 30 ns. The fuel in the target is thereby accelerated toward the centre by the pusher layer; the final implosion velocity reaches $1.9 \times 10^5$ m/s. After compression of the fuel, the ion temperature $T_i$ rises to 4.2 keV and $\rho R$ becomes 7.0 g/cm$^2$. The burn fraction of the fuel will be 34 %, and a fusion output thermal energy of 2.5 GJ will be released from a target.

Aluminium wires are connected to the target, through which a biasing voltage of $-1$ MV is applied to it to assist in focusing the proton beams on the target surface. Although the aluminium wires disturb the spherical symmetry of the target, the wires make it easier to set the target at the centre of the reactor cavity mechanically, with a repetition rate of 1 Hz.

The container of the reactor consists of double vessels, the inner one rotating around the horizontal axis while the outer one is fixed. Each vessel has six injector ports for the proton beams, and the beams are projected from power-supply systems outside the reactor container, synchronised with the port connections.

The characteristics of the power-supply systems from which the proton beams are projected are summarised in Table 4.8 (cf. Table 3.1). A power-supply system of an output voltage of 10 MV has already been designed in Japan, and is operating in several other countries. With the development of the plasma-erosion opening switch, however, the parameters given in section 3.6.2 are reasonable for a power-supply system of type 1. A power-supply system of type 3 is used to give the biasing voltage to the target.

If the gas switches in these systems are triggered by laser light, and water switches are replaced by magnetic switches, the jitter of the switches becomes less than 2 ns and the power-supply system can be operated synchronously.

A sketch of the fusion-reactor system is given in Fig. 4.25. As Fig. 4.23 shows, the six proton beams are projected into a cylindrical reactor cavity. Each beam is supplied by power-supply systems of type 1 and 2, which are located outside of the reactor cavity. An outline diagram of a type-1 power-supply system is given in Fig. 3.46. Another power-supply system is provided to maintain the target at a negative potential, in order to focus all the beam particles on the target surface. The evacuated gas and flibe from the reactor cavity are sent to the heat exchanger, exchanging their heat with $NaF-BF_3$. The $NaF-BF_3$ exchanges the heat again, this time to the water vapour which drives the turbines generating the electrical

**Fig. 4.25.** Power plant. **1** Power-supply system of type 1 and 2, **2** power supply system of type 3 for biassing target, **3** reactor cavity, **4** motor to rotate the reactor vessel, **5** D separator from the argon gas and the flibe, **6** T separator of the argon gas and the flibe, **7** heat exchanger from the flibe to $NaF-BF_3$, **8** heat exchanger from $NaF-BF_3$ to the water, **9** steam turbine, **10** electric power generator.

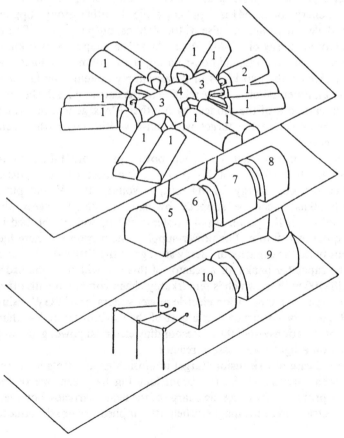

power, as described before. In addition, however, the first heat exchanger is connected with the system in which the unburnt D and T are separated from the inert gas. The bred T is also separated there from the flibe. The separated D and T are sent to the target-fabrication section to be filled into targets and used as the fuel.

These power-supply systems, including Marx generators, are so heavy that they are fixed to the floor. Since it is desirable that the six beams impinge on the target in an approximately spherically-symmetric way, the diodes and beam ports are distributed at spherically-symmetric positions on the cavity wall. Thus the length of the magnetically-insulated transmission lines between the power-supply systems and the diodes are not all equal. The firing times of the Marx generators must therefore be controlled so that the synchronised beams impinge simultaneously on the target. One power-supply system consists of two groups. From one group, pulsed electric power of pulse width 30 ns, output voltage 10 MV, and stored energy of 2.7 MJ is supplied, while the other group supplies pulsed electric power of pulse width of 30 ns, output voltage 5 MV, and stored energy of 1.7 MJ. As explained in chapter 3, two kinds of ion beams, rotating and non-rotating, are combined to make one beam. Hence the two kinds of power-supply system. The target is kept at a potential of $-1$ MV relative to the earthing level (the level of the cathodes of the diodes) by another Marx generator, which supplies power to the target with pulse width 200 ns (the beam propagation time).

The major parameters of such power plant are tabulated in Table 4.9. It is a feature of this kind of fusion reactor that its pulsed power – stored energy 2.7 MJ, output voltage 10 MV and pulse width 30 ns – can be supplied by capacity banks. Hence the technology of the power-supply systems will be well-established in the quite near future. The power-supply system proposed here has about twice the capacity of energy storage as the PBFA-II. The cost of the capacitor banks for a reactor of this size will be of the order of US$100 million, which is remarkably cheap compared with other fusion reactors. One fusion electric power generator of 1 GW, using LIB, could be constructed at a cost of the same order or less than the construction cost of the corresponding electrical power generator based on a light-water fission reactor.

In the case of LIB fusion, target implosion experiments have not yet been carried out. So far, beam focusing has been one of the major problems. By using the charge of the beam particles, however, the electric field or the magnetic field (the applied one or self-induced

Table 4.9 *Major parameters of power plant*

| | |
|---|---|
| Cavity shape | Cylinder |
| Cavity radius | 4 m |
| Total driver energy on a pellet | 7.9 MJ/shot |
| Pellet gain | 310 |
| Fractional burnup of fuel | 34 % |
| Fusion yield | 2500 MJ/shot |
| Repetition rate | 1 Hz |
| Fusion power | 2500 $MW_{th}$ |
| Driver efficiency | 25$\alpha$ |
| Ion type | Proton (9 MeV and 5 MeV) |
| Number of drivers | 6 |
| Cavity gas | Ar (0.1 Torr) |
| Maximum neutron wall loading | 30 $MW/m^2$ |
| T-breeding ratio | 1.1 |
| First wall and blanket structure | Flibe in Inconel 713C |
| Blanket breeding and heat transport medium | Flibe |
| Flibe inlet temperature | 750°C |
| Flibe outlet temperature | 550°C |
| Total thermal power | 2700 $MW_{th}$ |
| Gross electrical output | 879 $MW_e$ |
| Net electrical output | 800 $MW_e$ |
| Net plant efficiency | 33 % |

one) can confine the beam within a small radius. The advantages offered by the mechanism of beam stopping in the target of LIB, and the large amount of the total beam energy, show promise that sufficient fusion output energy will be released from the target.

The principles of the waterfall-type reactor are not proven yet but such a reactor does have the great merit of protecting the solid materials in the first wall from induced radio-activity. The solid wall in the solase-type reactor and the wet wall in the Hiball reactor can thus be applied to the LIB fusion reactor too.

## 4.4. Natural-resource requirements and safety of the fusion power reactor

Energy sources of the future must not only supply a sufficient amount of energy for human societies, but also the level of contamination of the environment they cause must be negligible. Does the energy released by the nuclear fusion satisfy these requirements? In this section, these requirements are examined.

### 4.4.1. Natural resources required
#### Fuel resources

The fuels for the D–T fusion power reactor are deuterium (D) and lithium (Li), as discussed in section 4.1.4. In Table 4.10, the required amounts of these resources are listed, together with those for other sources. Here Q is the unit used for a large amount of energy; $1 Q = 10^{18}$ BTU (British Thermal Unit) $= 1.055 \times 10^{21}$ J. World energy consumption is currently about 0.3 Q per year.

Deuterium is found in seawater at a concentration of 158 ppm, and in other waters at 140 ppm. To separate and collect D from water is quite easy; there are plants now working which extract D from the freshwater near estuaries. The small difference in the concentrations of D in the different kinds of water makes a large difference to the price of D. However, there are cases of concentrating plants that are now going out of commission due to corrosion of those parts of the plant which actually collect the D from seawater, because of the relatively high concentration of D in such water. So various technical problems remain to be solved before a large amount of D can be obtained at low cost.

Turning to the necessary Li resources, the amount of Li which will be consumed as fuel in a fusion power reactor will be very small. But as the blanket material for breeding T, 200–1000 tons of Li per 1 GW of electrical power output must be stored in a reactor.

The richest mines of Li are found in the USA, Canada and various African countries. The estimated amount of Li in all such mines has not been fixed, but seems to be of the order of $9 \times 10^6$ tons. A large amount of Li will be needed as the material for the batteries and for the coolant of a fast-breeder reactor. The $9 \times 10^6$ tons in Li mines will not be enough. However, there are salt lakes and spas whose waters include Li at high concentrations, 60–300 ppm, and these waters may supply sufficient Li. Investigation of Li as a natural resource has as yet been lacking, but with the development of research into fusion, such investigation will be increased in the near future.

Table 4.10 includes the natural resources of Li found in seawater. If we can utilise Li in seawater economically, we can solve the problem of fuel resources with respect to the D–T fusion reaction. The concentration of Li in the seawater is as low as 0.17 ppm, and it occurs along with Na, K, Ca, Mg, Br, etc. But it is thought that the Li can be separated without sophisticated technology because it is the smallest of the alkali metals found in seawater.

Table 4.10 Energy sources in the world (Q)

| Species of fuel | Known amount in mines | | | Estimated amount in mines |
|---|---|---|---|---|
| Fossil fuels | 23 | | | 1500 |
| Fuel for fission | | | | |
| Price($ /lb) | 5-10 | < 500 | > 500 | |
| Total energy | 490 | $1.6 \times 10^6$ | $5 \times 10^6$ | |
| No breeding | 7.2 | $2.6 \times 10^4$ | $7.5 \times 10^4$ | |
| (utilization rate 1.5% ) | | | | |
| Breeding | 300 | $9.6 \times 10^5$ | $3 \times 10^6$ | |
| Fuel for fusion | | | | |
| D | $7.5 \times 10^9$ | | | $7.5 \times 10^9$ |
| (total in the sea) | | | | |
| $_6$Li | $6 \times 10^6$ | | | $6 \times 10^9$ |
| (total in the sea) | | | | |

$1Q = 10^{18} BTU = 1.055 \times 10^{21} J$

### Resource materials other than fuel

The following materials will be required to construct a fusion power reactor:

*Structural materials:* Fe, Nb, Mo, V, Ni, Cr, Ti.
*T-breeders:* Li, flibe, $Li_2O$, $Al_2O_4$.
*Coolants:* He, Li, K, flibe.
*Superconducting materials:* NbTi, $Nb_3Sn$, $V_3Sn$, $V_3Ga$ (materials for wire); Cu, Al (materials for stabilisers); He (coolant).

Among these are some which on earth are very rare. In the design of the nuclear power reactor, account must be taken not only of the properties of a material but also the amount of that material required as a natural resource.

### 4.4.2. Safety of the nuclear fusion power reactor

It is very important to begin studying the safety of the fusion power reactor at the research stage, even though no fusion power reactor is currently operating. Much discussion and many evaluations of the safety of the fusion power reactor have already been conducted, and the results are reflected in the conceptual designs. In the present situation, it is not desirable to make decisive judgements on the safety of the various types of fusion reactors, but several questions may be asked and tentative answers given.

Safety problems may be classified into two groups, those that have to do with nuclei, and those that do not.

*Safety problems related to nuclei*

The main safety problems to do with nuclei relate to the tritium fuel and the radio-active structural materials irradiated by neutrons.

(i) *Tritium:* T is a radio-active material which undergoes $\beta$-decay and has a half-life of 12 years. The maximum energy of the $\beta$-ray is 18.6 keV and the average energy is 5.7 keV. Even with this weak radio-activity, T must be treated carefully because it diffuses easily in the form of waters (THO and $T_2O$). At ordinary temperatures, T is stored in a vessel, but at high temperatures the T will leak. If an animal drinks such waters, the waters remain in the body for a long time and it is known that they have particular effects on the sexual organs. From the viewpoint of safety, the amount of T stored in a vessel at high temperatures must be diminished. If the solid compounds of Li ($Li_2O$, $Li_2Al_2O_4$, and so on) can be used as the T-breeder, the amount of T stored in the blanket can be reduced.

Safety plans regarding T have been investigated from all possible sides. Concrete examples of basic treatment include the method of dividing the vessels in which T is enclosed into smaller ones so that the effect of an accident is limited to one or a few parts. Such multiple storage methods require that both multi-monitor systems and multi-withdrawal systems are also installed.

(ii) *Induced radio-activity:* In section 4.1.3, it was explained how the structural materials of the fusion power reactor induce radio-activity upon being irradiated by the neutron flux. The induced radio-active materials must then be scrapped. However, the radio-activity of these materials exceeds the lifetime of the reactor, and they also need to be cooled because they continue to release thermal energy during their decay. The fact that engineers cannot approach the reactor for maintenance and repairs because of the induced radio-activity is a major safety problem.

Table 4.11 shows the induced radio-activities of various structural materials; the values for 316SS, Nb–1Zr and V–20Ti have been obtained by calculation. The induced radio-activities depend on the material, so from the point of view of safety it is important that the materials are carefully chosen so that the half-lives of their induced radio-activities are not long, that the materials do not easily evaporate, and that they are not concentrated in the body of animals. As far as knowledge obtained from investigations so far conducted extends, Nb is the worst with respect to safety. It is also known that Al, V, SiC and C can be used to suppress induced radio-activities to low levels for cases of radio-activity with long half-lives.

Thermal energy released in the decay of radio-active nuclei is shown in Fig. 4.26. Even when Nb is used as the structural material,

Table 4.11 *Induced radio-activities of materials in the first wall for fusion power reactor* (operation over 10 years, wall loading 1.25 MW/cm$^2$)

| Structural material of the first wall | Induced radio-active materials* | Half-life | Radio-activity** (Ci/kW$_{th}$) |
|---|---|---|---|
| 316SS | $^{55}$Fe | 2.94 years | 140 |
| | $^{58}$Co | 72 years | 29 |
| | $^{54}$Mn | 310 days | 24 |
| | $^{60}$Co | 525 years | 47 |
| | | | Total*** 310 |
| Nb | $^{92}$Nb | 10.1 days | 152 |
| | $^{95m}$Nb | 3.8 days | 50 |
| | $^{95}$Nb | 35 days | 42 |
| | $^{89}$Sr | 51 days | 30 |
| | $^{94}$Nb | 50 000 years | 0.008 |
| | | | Total*** 290 |
| V | $^{48}$Sc | 1.8 days | 2.5 |
| | | | Total*** 5.6 |
| Al | $^{26}$Al | $7.5 \times 10^5$ years | 0.004 |
| | | | Total*** 0.004 |

\* Main atoms whose half-lives are more than 1 day
\*\* The total radio-activity in the blanket is 1.5–3 times the value in the table
\*\*\* Sum of radio-activities whose half-lives are less than 1 day

**Fig. 4.26.** Heat released by decay of nuclei after closing down the operation of the reactor

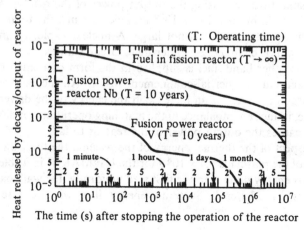

The time (s) after stopping the operation of the reactor

Table 4.12 *Stored energy in a D–T fusion power reactor with thermal output power of 5 GW*

| Mode of energy storage | |
| --- | --- |
| Plasma at the reactor centre (thermal energy) | $10^9$ J |
| Li for T-breeding (chemical energy) | $10^{12}$ J |
| Superconducting magnet (magnetic energy) | $10^{11}$ J |

the thermal energy released by decaying nuclei is only a few hundredths of the total power of the reactor, and in the case of V, it is less than one-thousandth. By comparison with a fission reactor, the thermal energy released by decaying nuclei in a fusion reactor will not be so severe a problem.

### Safety problems not related to nuclei

A large amount of energy is stored in a fusion power reactor of thermal output power 5 GW as shown in Table 4.12. Let us now look at the safety problems related to such energy.

(i) *Plasma at the reactor centre:* The first safety problem is the nuclear explosion of the plasma. In the case of magnetically-confined plasma, since the plasma pressure is balanced by the magnetic pressure, the nuclear reaction rate of the plasma particles cannot increase in some particular region. If the number density or the ion temperature is increased locally or momentarily by an instability, for instance, the increase will not spread to all the plasma and will not be sustained for a long time. An accident usually decreases the confinement time of the plasma, or induces mixing of impurities with the plasma, thus decreasing the output power of the reactor. With such a low plasma density as $10^{20}$ particles per m$^3$, the total mass of the fuel in the reactor is not large. A nuclear explosion in the plasma cannot be imagined under these circumstances.

On the other hand, fuel is supplied in the form of a small pellet to a reactor using inertial-confinement fusion. An increase of the fuel mass in a target, or an increase in the energy of the driver, for instance, induce an imbalance in the matching operation conditions, and so causes the output power of the reactor to fall.

In respect of the thermal energy of the plasma, the total thermal energy of the plasma is $6.4 \times 10^8$ J for the UWMAK-II in Table 4.4. If all this energy diffuses to the first wall uniformly, the wall loading increases by 25 J/cm$^2$. Such an energy flux will increase the temperature of the first wall, whose thickness is 1 cm, only a few degrees. However, if this energy flux is concentrated onto a small area of the first wall, the wall material will melt or evaporate. Such an accident can then trigger larger accidents. Accordingly, the blanket of the first wall is divided into many small units which can be replaced with new ones.

(ii) *Lithium:* Li is chemically very active: it burns when it is exposed to air, and reacts violently with many materials. Technologies enabling safe use of Li will be developed in the near future. The technology of using Na for cooling in the fast-breeder fission reactor has provided good suggestions for the technology of using Li in the fusion reactor.

In recent designs of fusion power reactors, however, chemically-inactive compounds, e.g. molten salts such as $Li_2BeF_4$, or solid compounds such as $Li_2O$, $Li_2Al_2O_4$ (see Table 4.4) are used instead of pure Li. With these compounds, the safety problems to do with using Li in the fusion reactor disappear.

(iii) *Superconducting magnets:* A large amount of magnetic energy and of liquid He are involved in a superconducting magnet. If the magnet loses superconductivity and reverts to functioning as a conventional magnet, the superconducting material is spoiled by the heat and the liquid He will boil.

A superconducting magnet is expensive and it takes time to cool the magnet sufficiently to recover superconductivity once the magnet has warmed to room temperature. Accidents to the superconducting magnet must therefore be avoided if at all possible. Although the reliability of the superconducting magnet has been increased through stabilising techniques, the technology for early detection of an unusual state of the superconductor and so of safeguarding the superconducting magnet remains to be developed. The following methods of safeguarding the magnet have been proposed:

(i) the energy stored in the magnet is extracted from the cryostat through a circuit within the time interval during which an abnormal voltage and heating above the limit do not occur;

(ii) each coil is divided into a number of sections to provide symmetry for the input and to hold the currents in balance. Using this method of magnetising the coil, the electro-magnetic forces induced by the superconducting magnets are in balance when the current in the magnet decreases. To deal with boiling of He, the reactor will be fitted with a breakable cover.

Although their size is not comparable with the superconducting magnet needed for a fusion reactor, useful knowledge has been accumulated from the operation of smaller ones. The technology for safeguarding large magnets has thus been developed.

•

# Appendix

SI (MKS) units are used except for temperature $T$ ($T$ in eV).

## Vector identities

Notation: $f, g$, etc., are scalars; $\mathbf{A}, \mathbf{B}$, etc., are vectors; $\mathbf{T}$ is a tensor

$$\mathbf{A} \cdot \mathbf{B} \times \mathbf{C} = \mathbf{A} \times \mathbf{B} \cdot \mathbf{C} = \mathbf{B} \cdot \mathbf{C} \times \mathbf{A} = \mathbf{B} \times \mathbf{C} \cdot \mathbf{A} = \mathbf{C} \cdot \mathbf{A} \times \mathbf{B} = \mathbf{C} \times \mathbf{A} \cdot \mathbf{B}$$

$$\mathbf{A} \times (\mathbf{B} \times \mathbf{C}) = (\mathbf{C} \times \mathbf{B}) \times \mathbf{A} = (\mathbf{A} \cdot \mathbf{C})\mathbf{B} - (\mathbf{A} \cdot \mathbf{B})\mathbf{C}$$

$$\mathbf{A} \times (\mathbf{B} \times \mathbf{C}) + \mathbf{B} \times (\mathbf{C} \times \mathbf{A}) + \mathbf{C} \times (\mathbf{A} \times \mathbf{B}) = 0$$

$$(\mathbf{A} \times \mathbf{B}) \cdot (\mathbf{C} \times \mathbf{D}) = (\mathbf{A} \cdot \mathbf{C})(\mathbf{B} \cdot \mathbf{D}) - (\mathbf{A} \cdot \mathbf{D})(\mathbf{B} \cdot \mathbf{C})$$

$$(\mathbf{A} \times \mathbf{B}) \times (\mathbf{C} \times \mathbf{D}) = (\mathbf{A} \times \mathbf{B} \cdot \mathbf{D})\mathbf{C} - (\mathbf{A} \times \mathbf{B} \cdot \mathbf{C})\mathbf{D}$$

$$\nabla(fg) = \nabla(gf) = f\nabla g + g\nabla f$$

$$\nabla \cdot (f\mathbf{A}) = f\nabla \cdot \mathbf{A} + \mathbf{A} \cdot \nabla f$$

$$\nabla \times (f\mathbf{A}) = f\nabla \times \mathbf{A} + \nabla f \times \mathbf{A}$$

$$\nabla \cdot (\mathbf{A} \times \mathbf{B}) = \mathbf{B} \cdot \nabla \times \mathbf{A} - \mathbf{A} \cdot \nabla \times \mathbf{B}$$

$$\nabla \times (\mathbf{A} \times \mathbf{B}) = \mathbf{A}(\nabla \cdot \mathbf{B}) - \mathbf{B}(\nabla \cdot \mathbf{A}) + (\mathbf{B} \cdot \nabla)\mathbf{A} - (\mathbf{A} \cdot \nabla)\mathbf{B}$$

$$\mathbf{A} \times (\nabla \times \mathbf{B}) = (\nabla \mathbf{B}) \cdot \mathbf{A} - (\mathbf{A} \cdot \nabla)\mathbf{B}$$

$$\nabla(\mathbf{A} \cdot \mathbf{B}) = \mathbf{A} \times (\nabla \times \mathbf{B}) + \mathbf{B} \times (\nabla \times \mathbf{A}) + (\mathbf{A} \cdot \nabla)\mathbf{B} + (\mathbf{B} \cdot \nabla)\mathbf{A}$$

$$\nabla^2 f = \nabla \cdot \nabla f$$

$$\nabla^2 \mathbf{A} = \nabla(\nabla \cdot \mathbf{A}) - \nabla \times \nabla \times \mathbf{A}$$

$$\nabla \times \nabla f = 0$$

$$\nabla \cdot \nabla \times \mathbf{A} = 0.$$

If $\mathbf{e}_1, \mathbf{e}_2, \mathbf{e}_3$ are orthonormal unit vectors, a second-order tensor $\mathbf{T}$ can be written in the dyadic form

$$\mathbf{T} = \sum_{i,j} T_{ij}\mathbf{e}_i\mathbf{e}_j.$$

In cartesian coordinates the divergence of a tensor is a vector with components

$$(\nabla \cdot \mathbf{T})_i = {\sum}'(\partial \mathbf{T}_{ji}/\partial x_j).$$

(This definition is required for consistency with eq. (1.28).) In general,

$$\nabla \cdot (\mathbf{AB}) = (\nabla \cdot \mathbf{A})\mathbf{B} + (\mathbf{A} \cdot \nabla)\mathbf{B}$$

$$\nabla \cdot (f\mathbf{T}) = \nabla f \cdot \mathbf{T} + f \nabla \cdot \mathbf{T}.$$

Let $\mathbf{r} = \mathbf{i}x + \mathbf{j}y + \mathbf{k}z$ be the radius vector of magnitude $r$ from the origin to the point $x, y, z$. Then

$$\nabla \cdot \mathbf{r} = 3$$

$$\nabla \times \mathbf{r} = 0$$

$$\nabla r = \mathbf{r}/r$$

$$\nabla(1/r) = -\mathbf{r}/r^3$$

$$\nabla \cdot (\mathbf{r}/r^3) = 4\pi\delta(\mathbf{r}).$$

If $V$ is a volume enclosed by a surface $S$ and $d\mathbf{S} = \mathbf{n}\, dS$ where $\mathbf{n}$ is the unit normal outward from $V$,

$$\int_V dV \nabla f = \int_S d\mathbf{S}\, f$$

$$\int_V dV \nabla \cdot \mathbf{A} = \int_S d\mathbf{S} \cdot \mathbf{A}$$

$$\int_V dV \nabla \cdot \mathbf{T} = \int_S d\mathbf{S} \cdot \mathbf{T}$$

$$\int_V dV \nabla \times \mathbf{A} = \int_S d\mathbf{S} \times \mathbf{A}$$

$$\int_V dV (f\nabla^2 g - g\nabla^2 f) = \int_S d\mathbf{S} \cdot (f\nabla g - g\nabla f)$$

$$\int_V dV (\mathbf{A} \cdot \nabla \times \nabla \times \mathbf{B} - \mathbf{B} \cdot \nabla \times \nabla \times \mathbf{A}) = \int_S d\mathbf{S} \cdot (\mathbf{B} \times \nabla \times \mathbf{A} - \mathbf{A} \times \nabla \times \mathbf{B}).$$

If $S$ is an open surface bounded by the contour $C$ of which the line element is $d\mathbf{l}$,

$$\int_S d\mathbf{S} \times \nabla f = \oint_C d\mathbf{l}\, f$$

$$\int_S d\mathbf{S}\cdot\nabla\times\mathbf{A} = \oint_C d\mathbf{l}\cdot\mathbf{A}$$

$$\int_S (d\mathbf{S}\times\nabla)\times\mathbf{A} = \oint_C d\mathbf{l}\times\mathbf{A}$$

$$\int_S d\mathbf{S}\cdot(\nabla f\times\nabla g) = \oint_C f\,dg = -\oint_C g\,df.$$

## Differential operators in curvilinear coordinates
### *Cylindrical coordinates*

Divergence

$$\nabla\cdot\mathbf{A} = \frac{1}{r}\frac{\partial}{\partial r}(rA_r) + \frac{1}{r}\frac{\partial A_\phi}{\partial \phi} + \frac{\partial A_z}{\partial z}.$$

Gradient

$$(\nabla f)_r = \frac{\partial f}{\partial r}; \qquad (\nabla f)_\phi = \frac{1}{r}\frac{\partial f}{\partial \phi}; \qquad (\nabla f)_z = \frac{\partial f}{\partial z}.$$

Curl

$$(\nabla\times\mathbf{A})_r = \frac{1}{r}\frac{\partial A_z}{\partial \phi} - \frac{\partial A_\phi}{\partial z}$$

$$(\nabla\times\mathbf{A})_\phi = \frac{\partial A_r}{\partial z} - \frac{\partial A_z}{\partial r}$$

$$(\nabla\times\mathbf{A})_z = \frac{1}{r}\frac{\partial}{\partial r}(rA_\phi) - \frac{1}{r}\frac{\partial A_r}{\partial \phi}.$$

Laplacian

$$\nabla^2 f = \frac{1}{r}\frac{\partial}{\partial r}\left(r\frac{\partial f}{\partial r}\right) + \frac{1}{r^2}\frac{\partial^2 f}{\partial \phi^2} + \frac{\partial^2 f}{\partial z^2}.$$

Laplacian of a vector

$$(\nabla^2\mathbf{A})_r = \nabla^2 A_r - \frac{2}{r^2}\frac{\partial A_\phi}{\partial \phi} - \frac{A_r}{r^2}$$

$$(\nabla^2\mathbf{A})_\phi = \nabla^2 A_\phi + \frac{2}{r^2}\frac{\partial A_r}{\partial \phi} - \frac{A_\phi}{r^2}$$

$$(\nabla^2\mathbf{A})_z = \nabla^2 A_z.$$

Components of $(\mathbf{A}\cdot\nabla)\mathbf{B}$

$$(\mathbf{A}\cdot\nabla\mathbf{B})_r = A_r\frac{\partial B_r}{\partial r} + \frac{A_\phi}{r}\frac{\partial B_r}{\partial\phi} + A_z\frac{\partial B_r}{\partial z} - \frac{A_\phi B_\phi}{r}$$

$$(\mathbf{A}\cdot\nabla\mathbf{B})_\phi = A_r\frac{\partial B_\phi}{\partial r} + \frac{A_\phi}{r}\frac{\partial B_\phi}{\partial z} + A_z\frac{\partial B_\phi}{\partial z} + \frac{A_\phi B_r}{r}$$

$$(\mathbf{A}\cdot\nabla\mathbf{B})_z = A_r\frac{\partial B_z}{\partial r} + \frac{A_\phi}{r}\frac{\partial B_z}{\partial\phi} + A_z\frac{\partial B_z}{\partial z}.$$

Divergence of a tensor

$$(\nabla\cdot\mathbf{T})_r = \frac{1}{r}\frac{\partial}{\partial r}(rT_{rr}) + \frac{1}{r}\frac{\partial}{\partial\phi}(T_{\phi r}) + \frac{\partial T_{zr}}{\partial z} - \frac{1}{r}T_{\phi\phi}$$

$$(\nabla\cdot\mathbf{T})_\phi = \frac{1}{r}\frac{\partial}{\partial r}(rT_{r\phi}) + \frac{1}{r}\frac{\partial T_{\phi\phi}}{\partial\phi} + \frac{\partial T_{z\phi}}{\partial z} + \frac{1}{r}T_{\phi r}$$

$$(\nabla\cdot\mathbf{T})_z = \frac{1}{r}\frac{\partial}{\partial r}(rT_{rz}) + \frac{1}{r}\frac{\partial T_{\phi z}}{\partial\phi} + \frac{\partial T_{zz}}{\partial z}.$$

### Spherical coordinates

Divergence
$$\nabla\cdot\mathbf{A} = \frac{1}{r^2}\frac{\partial}{\partial r}(r^2 A_r) + \frac{1}{r\sin\theta}\frac{\partial}{\partial\theta}(A_\theta\sin\theta) + \frac{1}{r\sin\theta}\frac{\partial A_\phi}{\partial\phi}.$$

Gradient
$$(\nabla f)_r = \frac{\partial f}{\partial r}; \qquad (\nabla f)_\theta = \frac{1}{r}\frac{\partial f}{\partial\theta}; \qquad (\nabla f)_\phi = \frac{1}{r\sin\theta}\frac{\partial f}{\partial\phi}.$$

Curl

$$(\nabla\times\mathbf{A})_r = \frac{1}{r\sin\theta}\frac{\partial}{\partial\theta}(A_\phi\sin\theta) - \frac{1}{r\sin\theta}\frac{\partial A_\theta}{\partial\phi}$$

$$(\nabla\times\mathbf{A})_\theta = \frac{1}{r\sin\theta}\frac{\partial A_r}{\partial\phi} - \frac{1}{r}\frac{\partial}{\partial r}(rA_\phi)$$

$$(\nabla\times\mathbf{A})_\phi = \frac{1}{r}\frac{\partial}{\partial r}(rA_\theta) - \frac{1}{r}\frac{\partial A_r}{\partial\theta}.$$

Laplacian

$$\nabla^2 f = \frac{1}{r^2}\frac{\partial}{\partial r}\left(r^2\frac{\partial f}{\partial r}\right) + \frac{1}{r^2\sin\theta}\frac{\partial}{\partial\theta}\left(\sin\theta\frac{\partial f}{\partial\theta}\right) + \frac{1}{r^2\sin^2\theta}\frac{\partial^2 f}{\partial\phi^2}.$$

Laplacian of a vector

$$(\nabla^2 \mathbf{A})_r = \nabla^2 A_r - \frac{2A_r}{r^2} - \frac{2}{r^2}\frac{\partial A_\theta}{\partial \theta} - \frac{2A_\theta \cot \theta}{r^2} - \frac{2}{r^2 \sin \theta}\frac{\partial A_\phi}{\partial \phi}$$

$$(\nabla^2 \mathbf{A})_\theta = \nabla^2 A_\theta + \frac{2}{r^2}\frac{\partial A_r}{\partial \theta} - \frac{A_\theta}{r^2 \sin^2 \theta} - \frac{2\cos\theta}{r^2 \sin^2 \theta}\frac{\partial A_\phi}{\partial \phi}$$

$$(\nabla^2 \mathbf{A})_\phi = \nabla^2 A_\phi - \frac{A_\phi}{r^2 \sin^2 \theta} + \frac{2}{r^2 \sin \theta}\frac{\partial A_r}{\partial \phi} + \frac{2\cos\theta}{r^2 \sin^2 \theta}\frac{\partial A_\theta}{\partial \phi}.$$

Components of $(\mathbf{A}\cdot\nabla)\mathbf{B}$

$$(\mathbf{A}\cdot\nabla\mathbf{B})_r = A_r\frac{\partial B_r}{\partial r} + \frac{A_\theta}{r}\frac{\partial B_r}{\partial \theta} + \frac{A_\phi}{r\sin\theta}\frac{\partial B_r}{\partial \phi} - \frac{A_\theta B_\theta + A_\phi B_\phi}{r}$$

$$(\mathbf{A}\cdot\nabla\mathbf{B})_\theta = A_r\frac{\partial B_\theta}{\partial r} + \frac{A_\theta}{r}\frac{\partial B_\theta}{\partial \theta} + \frac{A_\phi}{r\sin\theta}\frac{\partial B_\theta}{\partial \phi} + \frac{A_\theta B_r}{r} - \frac{A_\phi B_\phi \cot\theta}{r}$$

$$(\mathbf{A}\cdot\nabla\mathbf{B})_\phi = A_r\frac{\partial B_\phi}{\partial r} + \frac{A_\theta}{r}\frac{\partial B_\phi}{\partial \theta} + \frac{A_\phi}{r\sin\theta}\frac{\partial B_\phi}{\partial \phi} + \frac{A_\phi B_r}{r} + \frac{A_\phi B_\theta \cot\theta}{r}.$$

Divergence of a tensor

$$(\nabla\cdot\mathbf{T})_r = \frac{1}{r^2}\frac{\partial}{\partial r}(r^2 T_{rr}) + \frac{1}{r\sin\theta}\frac{\partial}{\partial\theta}(T_{\theta r}\sin\theta)$$
$$+ \frac{1}{r\sin\theta}\frac{\partial T_{\phi r}}{\partial\phi} - \frac{T_{\theta\theta}+T_{\phi\phi}}{r}$$

$$(\nabla\cdot\mathbf{T})_\theta = \frac{1}{r^2}\frac{\partial}{\partial r}(r^2 T_{r\theta}) + \frac{1}{r\sin\theta}\frac{\partial}{\partial\theta}(T_{\theta\theta}\sin\theta)$$
$$+ \frac{1}{r\sin\theta}\frac{\partial T_{\phi\theta}}{\partial\phi} + \frac{T_{\theta r}}{r} - \frac{\cot\theta}{r}T_{\phi\phi}$$

$$(\nabla\cdot\mathbf{T})_\phi = \frac{1}{r^2}\frac{\partial}{\partial r}(r^2 T_{r\phi}) + \frac{1}{r\sin\theta}\frac{\partial}{\partial\theta}(T_{\theta\phi}\sin\theta)$$
$$+ \frac{1}{r\sin\theta}\frac{\partial T_{\phi\phi}}{\partial\phi} + \frac{T_{\phi r}}{r} + \frac{\cot\theta}{r}T_{\phi\theta}.$$

## Other symbols and equations

Mass and energy

$$E = m_0 c^2 + \tfrac{1}{2}m_0 v^2$$
$$m_0 c^2 = 9 \times 10^{16}\text{ J} \qquad \text{for } m_0 = 1\,\text{kg}.$$

Rest mass and charge of elementary particle

| elementary particle | mass | charge |
|---|---|---|
| electron | $9.109 \times 10^{-31}$ kg | $-1.6 \times 10^{-19}$ C |
| proton | $1.673 \times 10^{-27}$ kg | $1.6 \times 10^{-19}$ C |
| neutron | $1.675 \times 10^{-27}$ kg | 0 |

Unit of energy

$$1\,\text{eV} = 1.602 \times 10^{-19}\,\text{J} = 11\,600\,\text{K}$$

$$1\,\text{Q} = 10^{18}\,\text{BTU} = 1.05 \times 10^{21}\,\text{J}.$$

Example of a fission reaction

$$^{235}_{92}\text{U} + ^{1}_{0}\text{n} = 2 \times ^{118}_{50}\text{Sn} + 8\text{e} + 236\,\text{MeV}.$$

Typical fusion reactions

$$4^{1}_{1}\text{H} + 2\text{e} = ^{4}_{2}\text{He} + 27.05\,\text{MeV}$$

$$^{2}_{1}\text{D} + ^{3}_{1}\text{T} = ^{4}_{2}\text{He} + ^{1}_{0}\text{n} + 17.6\,\text{MeV}$$

$$^{2}_{1}\text{D} + ^{2}_{1}\text{D} = \begin{cases} ^{3}_{2}\text{He} + ^{1}_{0}\text{n} + 3.27\,\text{MeV} \\ ^{3}_{1}\text{T} + ^{1}_{1}\text{H} + 4.03\,\text{MeV} \end{cases}$$

$$^{6}_{3}\text{Li} + ^{1}_{0}\text{n} = ^{4}_{2}\text{He} + ^{3}_{1}\text{T} + 4.8\,\text{MeV}$$

$$^{7}_{3}\text{Li} + ^{1}_{0}\text{n} = ^{4}_{2}\text{He} + ^{3}_{1}\text{T} + ^{1}_{0}\text{n} - 2.47\,\text{MeV}$$

$$^{9}_{4}\text{Be} + ^{1}_{0}\text{n} = 2^{4}_{2}\text{He} + 2^{1}_{0}\text{n}$$

$$^{11}_{5}\text{B} + ^{1}_{1}H = 3^{4}_{2}\text{He} + 8.7\,\text{MeV}$$

$$^{6}_{3}\text{Li} + ^{1}_{1}\text{H} = ^{3}_{2}\text{He} + ^{4}_{2}\text{He} + 4.0\,\text{MeV}.$$

Constants

| | |
|---|---|
| Boltzmann constant | $\kappa = 1.3709 \times 10^{-23}$ J/deg |
| Planck's constant | $h = 6.63 \times 10^{-34}$ J s |
| dielectric constant for a vacuum | $\varepsilon = 8.854 \times 10^{-12}$ F/m |
| magnetic permeability for a vacuum | $\mu = 4\pi \times 10^{-7}$ henry/m |

Maxwell distribution

$$f = n\left(\frac{m}{2\pi kT}\right)^{\frac{3}{2}} \exp\left\{-\frac{m}{2kT}v^2\right\}.$$

Coulomb collision cross-section

$$\sigma_m = \frac{1}{nl_m} = \frac{1}{nv\tau_m} = \frac{\langle \chi^2 \rangle}{nv} = 8\pi\rho_{r0}^2 \log \frac{r_D}{\rho_{r0}}$$

$$\rho_{r0} = \frac{Ze^2}{4\pi\varepsilon m_e v^2} = \frac{Ze^2}{12\pi\varepsilon kT}.$$

Mean free path

$$l = \frac{1}{\sigma n}.$$

Free-flight time

$$\tau_{ei} = \frac{25.8\pi^{\frac{1}{2}}\varepsilon^2 m_e^{\frac{1}{2}} T_e^{\frac{3}{2}}}{n_e Z e^4 \ln \Lambda} = 1.6 \times 10^{-10} \frac{1}{Z} T_e^{\frac{3}{2}} \left(\frac{n_e}{10^{20}}\right)^{-1}$$

$$\tau_{ii} = \frac{25.8\pi^{\frac{1}{2}}\varepsilon^2 m_i^{\frac{1}{2}} T_i^{\frac{3}{2}}}{n_i Z^4 e^4 \ln \Lambda} = 7.0 \times 10^{-9} \frac{A^{\frac{1}{2}}}{Z^4} T_i^{\frac{3}{2}} \left(\frac{n_i}{10^{20}}\right)^{-1}$$

$$\tau_{ie} = \frac{(2\pi)^{\frac{1}{2}} 3\pi\varepsilon^2 m_i T_e^{\frac{3}{2}}}{n_e Z^2 e^4 \ln \Lambda m_e^{\frac{1}{2}}} = 1.5 \times 10^{-7} \frac{A}{Z^2} T_e^{\frac{3}{2}} \left(\frac{n_e}{10^{20}}\right)^{-1}$$

Collision frequency

$$v_{ei} = \tau_{ei}^{-1} = 6.3 \times 10^9 Z T_e^{-\frac{3}{2}} \left(\frac{n_e}{10^{20}}\right)$$

$$v_{ii} = \tau_{ii}^{-1} = 1.4 \times 10^8 \frac{Z^4}{A^{\frac{1}{2}}} T_i^{-\frac{3}{2}} \left(\frac{n_i}{10^{20}}\right)$$

$$v_{ie} = \tau_{ie}^{-1} = 6.7 \times 10^6 \frac{Z^2}{A} T_e^{-\frac{3}{2}} \left(\frac{n_e}{10^{20}}\right).$$

Fusion reaction rate: for low energies ($T \leqslant 25$ keV) the data may be represented by

$$(\overline{\sigma v})_{DD} = 2.33 \times 10^{-14} T^{-\frac{2}{3}} \exp(-18.76 T^{-\frac{1}{3}}) \text{ cm}^3/\text{s}$$

$$(\overline{\sigma v})_{DT} = 3.68 \times 10^{-12} T^{-\frac{2}{3}} \exp(-19.94 T^{-\frac{1}{3}}) \text{ cm}^3/\text{s}$$

where $T$ is measured in keV.

Fusion output power

$$P_\alpha = \tfrac{1}{4}\langle \sigma v \rangle n^2 E_f = 8 \times 10^{-35} n^2 \text{ W/m}^3 \quad (10 \text{ keV, D–T, } E_f = 17.6 \text{ MeV}).$$

Bremsstrahlung

$$P_B = 5.35 \times 10^{-37} Z^2 n_e n_i T_e^{\frac{1}{2}} \text{ W/m}^3.$$

Lawson criterion

$$n\tau = \frac{24kT}{\langle\sigma v\rangle E_{\mathrm{f}} + 4.28 \times 10^{-36}T^{\frac{1}{2}}} = 10^{20} \text{ s/m}^3 \text{ (10 keV, D–T)}.$$

Cyclotron frequency

$$\text{electron}\qquad \omega_{\mathrm{ce}} = \frac{eB}{m_{\mathrm{e}}} = 1.76 \times 10^{11}B \text{ rad/s } (B; \text{T})$$

$$\text{proton}\qquad \omega_{\mathrm{cp}} = \frac{eB}{m_{\mathrm{p}}} = 0.96 \times 10^8 B \text{ rad/s } (B; \text{T}).$$

Gyroradius

$$\rho_{\mathrm{re}} = \left(\frac{T_{\mathrm{e}}}{m_{\mathrm{e}}}\right)^{\frac{1}{2}} \frac{1}{\Omega_{\mathrm{e}}} = \left(\frac{m_{\mathrm{e}}T_{\mathrm{e}}}{eB}\right)^{\frac{1}{2}} = 2.38 \times 10^{-6}T_{\mathrm{e}}^{\frac{1}{2}} \frac{1}{B}$$

$$\rho_{\mathrm{ri}} = \left(\frac{T_{\mathrm{i}}}{m_{\mathrm{i}}}\right)^{\frac{1}{2}} \frac{1}{\Omega_{\mathrm{i}}} = \left(\frac{m_{\mathrm{i}}T_{\mathrm{i}}}{eB}\right)^{\frac{1}{2}} = 1.02 \times 10^{-4} \frac{1}{Z} AT_{\mathrm{i}}^{\frac{1}{2}} \frac{1}{B}.$$

Plasma frequency

$$\omega_{\mathrm{p}} = \left(\frac{e^2 n_{\mathrm{e}}}{\varepsilon m_{\mathrm{e}}}\right)^{\frac{1}{2}} = 5.64 \times 10 n_{\mathrm{e}}^{\frac{1}{2}} \text{ rad/s}.$$

Debye radius

$$r_{\mathrm{D}} = \left(\frac{T_{\mathrm{e}}}{4\pi c^2 ne}\right)^{\frac{1}{2}} = 6.90 \times 10^3 \left(\frac{T_{\mathrm{e}}}{n}\right)^{\frac{1}{2}} \text{ m}.$$

Drift motion

electric field $\qquad\qquad\qquad\qquad\qquad v_{\mathrm{D}} = \dfrac{E_{\perp}}{B} \text{ m/s } (E_{\perp}; \text{V/m}, B; \text{T})$

gravity $\qquad\qquad\qquad\qquad\qquad\qquad v_{\mathrm{D}} = \dfrac{mg_{\perp}}{qB} = \dfrac{g_{\perp}}{\omega_{\mathrm{c}}}$

curvature of magnetic field $\qquad\qquad v_{\mathrm{D}} = \dfrac{mv_{\parallel}^2}{qBR}$

gradient of magnetic field intensity $\quad \dfrac{v_{\mathrm{D}}}{v_{\perp}} = \dfrac{\rho_{\mathrm{r}}|\nabla B|}{2B}$

divergence of magnetic field $\qquad\quad F_{\parallel} = -M_{\mathrm{m}}\dfrac{dB}{ds} = -\nabla_{\parallel}(\mathbf{M}_{\mathrm{m}}\mathbf{B}).$

Force exerted by combined electric and magnetic fields

$$\mathbf{F}' = \rho_e \mathbf{E} + \mathbf{j} \times \mathbf{B}.$$

Maxwell stress

$$P'_{ij} = \frac{1}{\mu} \left( \tfrac{1}{2} B^2 \delta_{ij} - B_i B_j \right).$$

Electric conductivity of plasma

$$C_e = \frac{\pi \varepsilon^2 (3kT_e)^{\frac{3}{2}}}{Ze^2 m_e^{\frac{1}{2}} \log \Lambda} = 5.2 \frac{T_e^{\frac{3}{2}}}{z \log \Lambda} \ \text{mho/m}.$$

Diffusion of plasma across magnetic field

classical
$$D_c = \frac{\rho_e^2}{\tau} \left( 1 + \frac{T_i}{T_e} \right)$$

Bohm
$$D_B = \frac{1}{16} \frac{kT_e}{eB}$$

neoclassical
$$D_{p-s} = D_c \left( 1 + \frac{8\pi^2}{\zeta^2} \frac{C_{e\perp}}{C_{e\parallel}} \right)$$

banana region
$$D_{ba} = \frac{q_s^2 \rho_i^2}{\varepsilon_a^{\frac{3}{2}} \tau}$$

plateau region
$$D_p = \frac{q_s \rho_i^2}{\varepsilon_a^{\frac{1}{2}} \tau_b}.$$

# References

## Chapter 1

1 A. Einstein: *Relativity, The Special and General Theory*, Henry Holt, New York (1920).
2 P. Putnam: *Energy in the Future* (1953).
3 A. B. Campbell: *Plasma Physics and Magnetohydrodynamics*, McGraw-Hill, New York (1963).
4 A. M. A. Leontovich: *Review of Plasma Physics*, vol. 1, Consultants Bureau, New York (1965).
5 G. Gamow: *Zeitschrift für Physik*, **55**, 204 (1924).
6 J. L. Tuck: *Nuclear Fusion*, **1**, 201 (1961).
7 W. Heitler: *The Quantum Theory of Radiation*, Oxford University Press, Oxford (1944).
8 H. R. Hulme: *Nuclear Fusion*, Bykeham Publications, London (1969).

## Chapter 2

9 F. F. Chen: *Introduction to Plasma Physics*, Plenum, New York (1974).
10 K. Miyamoto: *Plasma Physics for Nuclear Fusion*, The MIT Press, Cambridge, Massachusetts (1976).
11 S. Inoue, K. Itoh, T. Tange, K. Nishikawa and S. Yoshikawa: *Proc. 7th IAEA Conf.* (Innsbruck, 1978), CN-37/W-3.
12 D. Pfirsch and A. Schlüter: *Max-Planck-Institut Report*, MPI/PA/7/62.
13 A. A. Galeev and R. Z. Sagdeev: *Soviet Phys. JETP*, **26**, 233 (1968).
14 L. M. Kovrizhnykh: *Soviet Phys. JETP*, **29**, 475 (1969).
15 B. B. Kadomtsev and O. P. Pogutse: *Soviet Phys. JETP*, **24**, 1172 (1967).
16 T. Sato et al.: *Proc. 7th IAEA Conf.* (Innsbruck, 1978), CN-37/J-2.
17 Y. Tanaka, T. Tsunematsu, M. Azumi, S. Tokuda, G. Kurita and T. Takeda: *Proc. 11th Conf. on Numerical Simulation of Plasmas* (San Diego, 1983), 2B18.

## Chapter 3

18 H. Motz: *The Physics of Laser Fusion*, Academic Press, London (1979).
19 J. M. Duderstadt and G. A. Moses: *Inertial Confinement Fusion*, John Wiley & Sons, New York (1982).
20 W. Heitler: *The Quantum Theory of Radiation*, Oxford University Press, Oxford (1944).
21 Y. B. Zel'dovich and Y. P. Raizer: *Physics of Shock Wave and High-Temperature Hydrodynamic Phenomena*, Academic Press, New York (1966).
22 N. A. Krall and A. W. Trivelpiece: *Principles of Plasma Physics*, McGraw-Hill, New York (1973).
23 K. Nishikawa: *Advances in Plasma Physics*, vol. 6, John Wiley & Sons, New York (1976).
24 S. T. Craxton: *IEEE J. Quantum Electronics*, **QE-17**, 1981 (1981).
25 H. Azechi, M. Miyanaka, S. Sakabe, T. Yamanaka and C. Yamanaka: *Jpn J. Appl. Phys.*, **20**, L477 (1981).
26 P. L. Dreike, S. C. Glidden, J. B. Greenly, M. A. Greenspan, D. A. Hammer, S. Humphries, J. M. Neri, R. Pal, R. N. Sudan and L. G. Wiley: *Proc. 3rd Int. Topl. Con. High Power Electron and Ion Beam and Technology* (Novosibirsk, 1979).
27 P. F. Ottinger, S. A. Goldstein, D. Mosher and D. G. Colombant: *Proc. 5th Int. Topl. Conf. High Power Particle Beams* (San Francisco, 1983).
28 J. P. Van Devender, G. W. Barr, E. L. Burgess, D. L. Cook, A. M. Fine, J. P. Furans, A. A. Goldstein, R. A. Hamil, T. H. Martin, D. H. McDaniel, P. A. Miller, J. N. Olsen, K. R. Prestwich, J. Speldy, M. L. Tobyas and B. N. Jurman: *Proc. 6th Int. Conf. High Power Particle Beams* (Kobe, 1986).

## Chapter 4

29 E. S. Marmar: *J. Nucl. Mat.*, **76** and **77**, 59 (1978).
30 R. Behrish: *Nucl. Fusion*, **12**, 695 (1972).
31 W. Bauer: *J. Nucl. Mat.*, **76** and **77**, 9 (1978).
32 O. C. Yonts: *Proc. Nucl. Fusion Reactors Conf.* (Culham, 1969), p. 424.
33 G. M. McCracken: *Proc. 6th Int. Vacuum Congress* (Kyoto, 1974), p. 269.
34 M. Kaminsky and S. K. Das: *Int. Conf. Ion Surface Interaction-Sputtering* (Garching, 1972).
35 M. Kaminsky: *Proc. Int. Working Sessions on Fusion Reactors Tech.* (Oak Ridge, 1971), p. 86.
36 J. Bohdansky *et al.*: *J. Nucl. Mat.*, **76** and **77**, 163 (1978).
37 F. L. Vook *et al.*: *Rev. Mod. Phys.*, **47** (Suppl.), No. 3 (Winter, 1975).

38 S. Blow: *Proc. Int. Working Sessions on Fusion Reactors Tech.* (Oak Ridge, 1971), p. 46.
39 R. Toschi: *IEEE Transactions on Mag.*, **MAG-14**, 586 (1978).
40 *Status and Objective of Tokomak Systems for Fusion Research*, USAEC WASH 1295, III.B.1 and 3.
41 *The Jet Project*, EUR 5791 e, 175 (1977).
42 K. Sako et al.: *Proc. 5th Int. Conf. on Plasma Phys. and Controlled Nucl. Fusion Research*, vol. III (Tokyo, 1974), p. 535.
43 R. W. Conn et al.: *ibid.*, p. 497.
44 W. G. Price Jr: *ibid.*, p. 525.
45 R. W. Conn et al.: *Proc. 6th Int. Conf. on Plasma Phys. and Controlled Nucl. Fusion Research*, vol. III (Berchtesgaden, 1976), p. 203.
46 J. File et al.: *IEEE Transactions on Nucl. Sci.*, **NS-18**, 277 (1971).
47 R. W. Werner et al.: *Proc. IAEA Workshop Fusion Reactor Design Problems* (1974), p. 171.
48 R. F. Post: *Proc. Nucl. Fusion Reactors Conf.* (Culham, 1969), p. 88.
49 R. W. Moir et al.: *Proc. 4th Int. Conf. on Plasma Phys. and Controlled Nucl. Fusion Research*, vol. III (Madison, 1971), p. 315.
50 R. W. Moir and W. L. Barr: *Nucl. Fusion*, **13**, 35 (1973).
51 W. L. Barr et al.: *IEEE Transactions on Plasma Sci.*, **PS-2**, 71 (1974).
52 T. O. Hunter and G. L. Kulcinski: *J. Nucl. Mat.*, **76** and **77**, 383 (1978).
53 L. A. Booth et al.: *Proc. IEEE*, **64**, 1460 (1976).
54 T. Frank et al.: *Proc. 1st Tropical Meeting on the Technol. of Controlled Nucl. Fusion*, vol. I (San Diego, 1974), p. 83.
55 L. A. Booth: *Los Alamos Sci. Lab. Rep.*, LA-4858-MS, vol. I (1972).
56 A. P. Fraas: *Oak Ridge Natl Lab. Rep.*, ORNL-TM-3231 (1971).
57 J. A. Maniscalco et al.: *Technical Digest at the Topical Meeting on Inertial Confinement Fusion* (San Diego, 1978), paper WC 3.
58 M. J. Monsler and J. A. Maniscalco: *ibid.*, paper WC 5.
59 R. W. Conn et al.: *Rep. Univ. of Wisconsin*, UWFDM-220 (1977).
60 J. Nuckolls et al.: *Nature*, **239**, 138 (1972).
61 L. M. Frantz and J. S. Nodvik: *J. Appl. Phys.*, **34**, 2346 (1963).
62 A. J. Demaria: *Proc. IEEE*, **61**, 731 (1973).
63 O. R. Wood II: *ibid.*, **62**, 355 (1974).
64 B. J. Feldman: *IEEE J. Quantum Electronics*, **QE-9**, 1070 (1973).
65 G. Brederlow et al.: *ibid.*, **QE-12**, 152 (1976).
66 G. Emanuel et al.: *J. Quantum Spectrosc. Radiat. Transfer*, **13**, 1365 (1973).
67 J. H. Parker and G. C. Pimentel: *J. Chem. Phys.*, **51**, 91 (1969).
68 J. C. Polanyi and K. B. Woodall: *ibid.*, **57**, 1574 (1975).
69 J. L. Tuck: *Proc. Nucl. Fusion Reactors Conf.* (Culham, 1969), p. 31.

70 N. Walton and E. Spooner: *Nature*, **261**, 533 (1976).
71 G. L. Kulcinski: *Proc. 5th Int. Conf. on Plasma Phys. and Controlled Nucl. Fusion Research*, vol. II (Tokyo, 1974), p. 251.
72 D. Steiner and A. P. Fraas: *Nucl. Safety*, **13**, 353 (1972).
73 G. E. Goker and L. J. Perkins: *Fusion Tech.*, **8**, 332 (1985).
74 C. C. Baker, G. A. Carlson and R. A. Karakowski: *Nuclear Tech./Fusion*, **1**, 5 (1981).
75 S. Shimamoto: *Proc. 11th Int. Conf. Plasma Phys. and Controlled Nuclear Fusion Res.* (Kyoto, 1986).
76 H. Madarame, S. Iwaki, A. Susuki, Y. Ogata, Y. Oka and S. Tanaka: *Univ. of Tokyo Report*, UTNL-R-0144 (1982).
77 K. Niu and S. Kawata: *Fusion Tech.*, **11**, 1365 (1987).

# Index